Biotechnology

DRUGS AND THE PHARMACEUTICAL SCIENCES
A Series of Textbooks and Monographs

Series Executive Editor
James Swarbrick
PharmaceuTech, Inc.
Pinehurst, North Carolina

Recent Titles in Series

For more information about this series, please visit: www.crcpress.com/Drugs-and-the-Pharmaceutical-Sciences/book-series/IHCDRUPHASCI

Biotechnology

The Science, the Products, the Government, the Business

Ronald P. Evens

CRC Press
Taylor & Francis Group
Boca Raton London New York

CRC Press is an imprint of the
Taylor & Francis Group, an **informa** business

First edition published 2020
by CRC Press
6000 Broken Sound Parkway NW, Suite 300, Boca Raton, FL 33487-2742

and by CRC Press
2 Park Square, Milton Park, Abingdon, Oxon, OX14 4RN

© 2020 Taylor & Francis Group, LLC

CRC Press is an imprint of Taylor & Francis Group, LLC

ISBN: 978-0-367-02467-3 (hbk)
ISBN: 978-0-367-49702-6 (pbk)
ISBN: 978-0-429-39929-9 (ebk)

Library of Congress Cataloging-in-Publication Data

Names: Evens, Ronald P., author.
Title: Biotechnology : the science, the products, the government, the business / by Ronald P. Evens.
Description: First edition. | Boca Raton, FL : CRC Press, 2020. | Series: Drugs and the pharmaceutical sciences | Includes bibliographical references and index.
Identifiers: LCCN 2020009451 | ISBN 9780367024673 (hardback) | ISBN 9780429399299 (ebook) | ISBN 9780367497026 (paperback)
Subjects: LCSH: Pharmaceutical biotechnology. | Pharmaceutical biotechnology industry.
Classification: LCC RS380 .E94 2020 | DDC 615.1/9--dc23
LC record available at https://lccn.loc.gov/2020009451

Typeset in Palatino
by Lumina Datamatics Limited

Contents

Preface

We have observed that "Biotechnology" over the last 50 years has truly created and continues to result in a revolution – in scientific advances, in the biopharma industry with leadership in novel product development, and in health care advances with cures or substantial disease mitigation. In my 45-plus years in pharmaceutical practice, academia and industry, I have been fortunate to learn about and track biotechnology's science and business opportunities and advances and now can share my assimilated biotechnology information with my colleagues in academia and industry in this form of a book, *Biotechnology: The Science, the Products, the Government, the Business*. I have been very fortunate in my profession and career to have had manifold, broad, and challenging work environments and substantial growth opportunities. The intent of this book is for me to give back and to offer material (narratives and my data slides) to further enhance the educational opportunities in biotechnology and to assist in the mentoring and education of pharmacy, medical, and graduate students and young practitioners. I thank my family and mentors at the universities at Buffalo, Kentucky, and Texas for their advice and support, early in my education and career.

Author

Dr. Ronald P. Evens is Adjunct Research Professor and biotechnology consultant at Tufts University, School of Medicine, Center for the Study of Drug Development. Dr. Evens is President of MAPS 4 Biotec, Inc., his consulting company to biotechnology industry for strategy, planning, and operations of medical affairs groups for drug research, medical education, medical liaisons, provider/patient services, and product launch. Clients have included Amgen, Amylin, Bertek, Cheladerm, Oragenics, Pharmacyclics in addition to advising Banyan, Prevacus, MLM, Nanotherapeutics, and St. Charles. He is editor of two books, *Drug & Biological Development: From Molecule to Product & Beyond,* Springer Publisher, 2007; *BioPharma Language, Acronyms and Terms,* Jones and Bartlett Publisher, 2009. Dr. Evens has also authored 16 book chapters in biotechnology.

Dr. Evens was Professor of Pharmacy at the University of Florida and then the University of the Pacific for 15 years, with a focus on biotechnology education. At Amgen [worldwide leader in biotech industry (13 years)], he created and was Senior Director of the Professional Services Department (with a staff of 140 and a budget of more than $60 million), including clinical research, medical education, medical information, and executive management with eight products and seven groups, plus international coordination of medical affairs. He also created the PeriApproval Research group for pipeline products. He has participated in the redesign of Amgen's research and development process, the sales and marketing group, and the leadership development program. He was a Clinical Professor at the University of Southern California at that time. Before Amgen, Dr. Evens was Associate Director, Clinical Research and Medical Services at Bristol-Myers Co. (6 years), covering central nervous system drugs. Dr. Evens was an Associate Professor and acting Chairman of the Department of Pharmacy Practice at the University of Tennessee (3 years) and Associate Professor and Director of the Drug Information Center at the University of Texas at Austin and University of Texas Center for Health Sciences at San Antonio (7 years). Dr. Evens received a BS in Pharmacy at the University of Buffalo, New York (1969). His graduate work was at the University of Kentucky for both his PharmD and clinical practice residency (1971–1974).

Dr. Evens has served on 12 Boards of Directors or Advisory Boards for professional societies. He was on the Boards of Cheladerm and Oragenics companies, including interim Chief Executive Officer. He has given more than 150 scientific, professional, and industry presentations at national and state professional society meetings. His publications exceed 100, including 14 book chapters and editor of 2 books. The University of Buffalo recognized him with four awards: Roger Mantsavinos (Biochemistry), Robert Ritz (Pharmacology), Rexall (first in class), and Rho Chi Honor Society. The University of Kentucky awarded him the clinical residency "Impact Award." The American College of Clinical Pharmacy recognized him with their fellowship. Otherwise, Dr. Evens' interests are the environment, support of the disadvantaged in society, pharmacy and medical education, as well as blues music, history, travel, and sports and, most of all, his family (his loving partner and three adult children: two physicians and a botanist).

List of abbreviations

A	adenine
a.a.	amino acid
Ab	antibody
ACS	acute coronary syndrome
ADC	antibody drug conjugate
ADA	adenosine deaminase deficiency
Ag	antigen
ADCC	antibody dependent cell cytotoxicity
AMA	American Medical Association
ADME	absorption, distribution, metabolism, excretion
APC	antigen presenting cells
APD	antibody phage display
AS	ankylosing spondylitis
ASCPT	American Society of Clinical Pharmacology and Therapeutics
AZ	AstraZeneca company
B	billion
Bb	blockbuster
BCG	Bacillus Calmette-Guerin
BCPA	Best Childrens Pharmaceutical Act
BHK	baby hamster kidney
B-I	Boehringer-Ingelheim company or Biogen-Idec
BioSci	Biosciences
BLA	Biologics License Application
BMP	bone morphogenetic protein
BMS	Bristol Myers-Squibb company
BOD	board of directors
BTD	breakthrough therapy designation
C	cytosine or centigrate or cancer
Ca or caz	cancer
CAPS	cryopyrin-associated periodic syndrome
CAR-T	chimeric antigen receptor T-cells
CBER	Center for Biologics Evaluation & Research

CD	complementary determining (antigen)
CDC	Center for Disease Control
CDER	Center for Drug Evaluation & Research
cDNA	complementary deoxyribonucleic acid
CDCC	complement dependent cell cytotoxicity
CDR	complementary determining receptor
CEO	chief executive officer
CHO	carbohydrate or Chinese hamster ovary cells
CMC	chemistry manufacturing and controls
CMO	chief medical officer
CMS	Centers for Medicare and Medicaid Services
CMV	Cytomegalovirus
Co.	Company
Comp.	Company
Conc	concentration
COOH	carboxyl group
CPP	central precocious puberty
CPT	Clinical Pharmacology & Therapeutics
CRC	colorectal cancer
CRISPR	clustered regulatory interspaced short palindromic repeats
CRL	complete response letter from FDA
CSF	colony stimulating factor
CSO	chief science officer
CT	cell therapy
CTA	clinical trials application
CTD	clinical trials document
CV	cardiovascular
D	dermatology
Def	definition
Defic	deficiency
Derm.	Dermatology
D.ins.	diabetes insipidus
Dis	disease
DLCBL	diffuse large cell b-cell lymphoma
D.m.	diabetes mellitus
DNA	deoxyribonucleic acid
DOA	Department of Agriculture
DOD	Department of Defense
DOJ	Department of Justice
D-S	Daiichi-Sankyo company
DTC	direct-to-consumer advertising
Duch.	Musc. Dys. Duchenne's muscular dystrophy
DVT	deep vein thrombosis
EMA	European Medicines Agency

EMEA	European Medicines Evaluation Agency
Eng.	engineering
Endo	endocrinology
Eos	eosinophil
EPO	epoetin alfa
EU	European Union
F	factor
Fab	antibody fragment
FDA	Food & Drug Administration
FDAMA	Food & Drug Administration Modernization Act
FDASIA	Food & Drug Administration Safety and Innovation Act
FHT	familial hereditary transthyretin amyloidosis
FIPCO	fully integrated pharmaceutical company
FMF	Familial Mediterranean Fever
FP	fusion protein
FTC	Federal Trade Commission
G	guanine or generation
GCP	good clinical practices
GE	gastroenterology
Gen	genetics or Genentech company
GF	or g.f. growth factor
GI	gastrointestinal
GLP	good laboratory practices
GM	or Gm granulocyte-macrophage or gram
GMO	genetically modified organism
GMP	good manufacturing practices
GRP	good regulatory practices
GSK	GlaxoSmithKline company
Gyn	gynecology
H	hematology
HAMA	human anti-mouse antibody
HCP	health care professional
Hem	hematology
HGS	Human Genome Sciences company
HIV	human immunodeficiency virus
HMO	health maintenance organization
HPLC	high pressure liquid chromatography
HPV	human papilloma virus
HTS	high throughput screening
HUS	hemolytic uremic syndrome
IBS	Inflammatory bowel syndrome
ID	infectious disease or identification
IDE	investigational device exemption
IFN	interferon

Ig	immunoglobulin
IL	interleukin
IM	internal medicine or intramuscular
IND	Investigational New Drug application
Inf. dis.	infectious disease
IOM	Institute of Medicine
IPO	initial public offering
IR	investor relations
IRS	Internal Revenue Service
Jap.	Japan
JnJ	Johnson & Johnson company
Jour	journal
JRA	juvenile rheumatoid arthritis
Kg	kilogram
K-K	Kyowa Kirin company
L.a.	long acting
LAR	long acting release
Leuk.	leukemia
LH	luteinizing hormone
Lymph	lymphoma
M	million
M&A	merger & acquisitions
Mab	monoclonal antibody
MBA	master in business administration
MD	medical doctor
Mcg	microgram
MCWB	master cell working bank
Med	medical or medicine
MHC	major histocompatibility complex
miRNA	micro-inhibitory ribonucleic acid
MKD	mevalonate kinase deficiency
MOA	mechanism of action
Mol. eng.	molecular engineering
mRNA	messenger ribonucleic acid
MRSA	methicillin resistant staphylococcus aureus
N	number or neurology
n.a.	nucleic acid
NBE	new biological entity
NCI	National Cancer Institute
NDA	new drug application
Neuro	neurology
NH2	amino group
NHL	non-hodgkin's lymphoma
NIH	National Institutes of Health

NME	new molecular entity
No.	number
Noct	nocturnal
NSCLC	non-small cell lung cancer
ODD	orphan drug designation
OIG	Office of Inspector General
ON	Oncology
Onc.	oncology
Oph	Ophthalmology
Ops	operations
OV	ovarian cancer
PBPC	peripheral blood progenitor cells
P'col	pharmacology
PCR	polymerase chain reaction
PD	pharmacodynamics
PDCC	phagocyte dependent cell cytotoxicity
PDGF	platelet derived growth factor
PDUFA	Prescription Drug User Fee Act
PEG	polyethylene glycol
Pep	peptide
PhD	doctor of philosophy
PIPE	private investment in public equity
PK	pharmacokinetics
PMA	pre-marketing approval (devices)
PMC	post marketing commitments
PNH	paroxysmal nocturnal hemoglobinuria
POA	plan of action
POP	proof of principle
Post. op.	post-operative
PPM	pharmacy prescription management
PPO	physician practice organization
PR	public relations
Ps	Psoriasis
PsA	Psoriatic arthritis
P&T	Pharmacy & Therapeutics committee
Pulm	Pulmonary
QA	quality assurance
QC	quality control
QIDP	qualified infectious disease product
®	registered name
RA	rheumatoid arthritis
R&D	research and development
RCC	renal cell carcinoma
rDNA	recombinant deoxyribonucleic acid

Reg	regulatory
REMS	risk evaluation & mitigation strategies
Resp	respiratory
Rh	rheumatology
RISC	RNA induced silencing complex
RMAT	regenerative medicine advanced therapy designation
RNA	ribonucleic acid
RNAi	ribonucleic acid inhibition
RNase	ribonucleic acid enzyme
ROW	rest of world
rRNA	ribosomal ribonucleic acid
RSV	respiratory syncytial virus
Rx	prescription
SBIR	small business innovation research
SCF	stem cell factor
scFv	single chain variable fragment antibody
SCID	severe compromised immunologic disease
SCT	stem cell therapy
SEC	Security and Exchange Commission
S&M	sales and marketing
SMA	spinal muscle atrophy
sNDA	supplemental new drug application
SNP	single nucleotide polymorphisms
Syn	synthetic
Synd	syndrome
T	thymidine or time
TA or T.a.	therapeutic area
Tak	Takeda company
TALEN	transcription activator-like effector nuclease
Therap.	therapeutics
TNF	tumor necrosis factor
Tox	toxicology
tRNA	transfer ribonucleic acid
TR	transplant
TRAPS	tumor necrosis factor receptor associated periodic syndrome
Tx	treatment
U	uracil
UC	ulcerative colitis
USA	United States of America
USDA	United States Department of Agriculture
VA	Veterans Administration
Vac	vaccine
VC	venture capital
VOC	veno-occlusive crisis

WBC	white blood cell
XR	X-ray
Yrs	years
ZFN	zinc finger nuclease
351 (k)	marketing application with FDA for biosimilar product
510 (k)	premarket application with FDA for devices

chapter one

Introduction

The scientific revolution in drug discovery, product development, and patient care that occurred near the end of the twentieth century, that is, the advent and full realization of the value and applications of biotechnology in health care, continues unabated and even accelerated into the twenty-first century. The growth and advances in biotechnology have been remarkable in science (multiple novel technologies created and numerous new products approved), in disease management (myriad indications, great extent of disease mitigation, and even manifold cures), and in the business arena (extensive list of varied biotech companies, universal pharma companies' engagement, billion-dollar investments, and blockbuster sales). In the early era of biotech product approvals, we observed only 15 new products in the 1980s, which improved to 67 in the 1990s, over this 18-year pre-millennial period. After 2000, this discipline of biotechnology has experienced more than a quadrupling of the discovery, development, production, and commercialization of innovative biological products, with about 360 more products by the end of 2019 (a second 18-year period). Moreover, over 375 human disease conditions now are treated with the 442 biotechnology products, many of which continue to be major medical breakthroughs, which are marketed by about 100 biotech and pharma companies.

History of biotechnology

The earliest origins of biotechnology harken back thousands of years, employing the process of fermentation to create new food sources. About 8,000 to 4,000 years ago, beer, wine, cheeses, and breads were created by utilizing the rudimentary tools of grains (sources of carbohydrates and proteins to be manipulated into food stuffs), a yeast (a living organism with biochemical and enzymatic properties), and a container vessel in which the source materials, yeasts, and processes can function. Also, genetic manipulation of animal and plants has been performed over many millennia through crossbreeding to create improved traits in plants and animals as food sources and agricultural products for sale. Moldy soy was used in China to treat boils hundreds of years ago. Inoculation against smallpox was performed using extracts from smallpox lesions in the revolutionary British colonies in North America. Genetics began with the discoveries related to heredity by Charles Darwin and Gregor Mendel in the mid-1800s; the principles were used in breeding animals and plants to enhance desirable traits. The existence of proteins was discovered in the 1800s, and DNA being responsible for carrying genetic information was established in 1940s. The first biotech product, that was genetically created, manufactured, clinically studied, and approved for use, was a peptide hormone, recombinant human insulin, which was created in 1982 by the company Genentech. For medical biotechnology and product development, critical discoveries in the biological sciences especially over the preceding 30 years that led to this first product achievement are identified below in a timeline (Table 1.1).

Table 1.1 Timeline for biologic discoveries leading to biotech product development

---

1663 – Robert Hooke first described living cells.

1677 – Antonie van Leeuwenhoek discovered bacteria and protozoa.

1798 – Edward Jenner was the first to use a viral vaccine as an inoculation for smallpox.

1838 – Gerardus Mulder and Jons Berzeluis discovered and named proteins.

1862 – Louis Pasteur discovered fermentation from bacterial origin.

1863 – Gregor Mendel determined the laws of inheritance.

1869 – Freidrich Miescher identified DNA in trout sperm.

1878 – Walther Flemming discovered chromatin, which lead to the discovery of chromosomes.

1881 – Louis Pasteur created vaccines against bacterial infections, e.g., anthrax.

1888 – Heinrich von Waldeyer-Hartz discovered chromosomes.

1909 – Genes were first linked with hereditary diseases.

1915 – Phages, bacterial viruses, were discovered.

1919 – The word, "biotechnology," was first used by a Hungarian agricultural engineer, Karl Ereky.

1953 – Drs. James Watson and Francis Crick developed the three-dimensional configuration of DNA with the paired nucleotides and a double helical shape.

1955 – The enzyme, DNA polymerase, responsible for synthesis of nucleic acid-based molecules, was isolated.

1959 – Gerald Edelman and Rodney Robert Porter independently discovered the structure of antibodies.

1960 – Messenger RNA was discovered to carry the genetic code from the nucleus.

1961 – Dr. Marshall Nirenberg established that genetic information in DNA determines protein structure and is related to most cellular functions.

(Continued)

Table 1.1 (*Continued*) Timeline for biologic discoveries leading to biotech product development

---

1961–1965 – Genetic code was delineated, responsible for translating the connection between DNA and proteins:
- Genetic code (DNA and RNA) was revealed to be generally universal in all living things.
- In the genetic code, the four RNA nucleotides (two purines – adenine and guanine and two pyrimidines – uracil and cytosine) in varied triplet codes (over 60 triplets) had a direct relationship to the 20 amino acids in determining the structure of proteins.
- The multifaceted and interrelated roles of several RNA species, that is, messenger RNA, ribosomal RNA, and transfer RNA, were elucidated in the transcription and translation processes, resulting in nuclear DNA codes being transformed into the production of proteins in cell cytoplasm.

1963 – The peptide insulin was synthesized.

1964 – Reverse transcriptase enzyme was discovered for creating in reverse order, DNA from RNA.

1970 – Hamilton Smith discovered restriction enzyme, responsible for cutting DNA into fragments, permitting genetic manipulation.

1971 – The biotech company Cetus was founded.

1972 – DNA ligase was used to recombine DNA fragments, further permitting genetic manipulation.

1972 – Dr. Paul Berg combined genes isolated from different organisms (mammalian or bacteria) into a hybrid DNA molecule.

1973 – Drs. Stanley Cohen (Stanford University) and Herbert Boyer (University of California, Berkley) created the first recombinant DNA experiment, inserting a DNA molecule (fragment) into another cell's DNA with replication/cloning of DNA.

1974 – The National Institutes of Health (NIH) in the United States formed the Recombinant DNA Advisory Committee for oversight of recombinant genetic research.

1975 – Monoclonal antibody technology was developed by Drs. Georges Kohler, Neils Jerne, and Cesar Milstein through hybridoma cell development, that produce antibodies versus specific antigens and demonstrating hybridomas are immortal.

1976 – The biotech company Genentech was founded.

1977 – DNA sequencing techniques were developed, permitting accurate sequencing of genes, separately by both Dr. Walter Gilbert in the United States and Dr. Frederick Sanger in England.

1977 – The human peptide for growth hormone, somatostatin, was produced in a genetically engineered bacteria, a first for cloning of a human protein or peptide with a human gene inserted in a nonhuman organism's DNA.

(Continued)

Table 1.1 (*Continued*) Timeline for biologic discoveries leading to biotech product development

1978 – Dr. Herbert Boyer created the first biologically synthetic version of human insulin, inserting a human insulin gene into the bacterium *Escherichia coli*, producing the first human peptide.

1978 – Antisense was described by Paul Zamecnik and Mary Stephenson.

1978 – The biotech companies Biogen, Agouron, and Hybritech were founded.

1979 – The biotech company Centocor was founded.

1980 – The U.S. Supreme Court decided that genetically altered life forms can be patented, creating tremendous commercial opportunities for genetic engineering.

1980 – The biotech companies Amgen, Genetics Institute, and Genetic Systems were founded.

1980s – Transgenic animals were first created in various livestock.

1981 – The biotech companies Chiron, Genzyme, and Immunex were founded.

1982 – The U.S. Food and Drug Administration (FDA) approved the first biotechnology product, recombinant human insulin, a peptide produced from genetic engineering, which was created by Genentech and licensed to Eli Lilly for sale in the treatment of diabetes mellitus.

1983 – Dr. Kay Banks Mullis invented the polymerase chain reaction technique to multiply DNA sequences and revolutionized molecular biology and genetic engineering.

1983 – The first genetically engineered plant was created by inserting an antibiotic resistance gene into a tobacco plant.

1986 – The first genetically engineered vaccine for hepatitis B prevention was created by Chiron and was approved by the FDA.

1986 – The first monoclonal antibody, Orthoclone OKT3, from biotechnology was approved by the FDA from Centocor as a therapy for rejection of kidney transplants.

1987 – The first biological (protein) was produced in a transgenic animal's milk (mouse).

1988 – The biotech company MedImmune was founded.

1990 – First gene therapy was successfully used to treat the immune disorder, severe compromised immune deficiency, which is an adenosine deaminase enzyme deficiency in human lymphocytes, leading to death from infections in children.

1990 – The international Human Genome Project was launched to map all the genes in the human genome.

1990 – The first fusion protein, Adamase (pegademase), was approved for use in severe compromised immune deficiency.

1990 – The recombinant protein, Pulmozyme (dornase alfa), to be given by inhalation instead of typical administration of proteins by injection, was approved for respiratory anomalies in cystic fibrosis.

1990 – Total approved biotech products by the end of 1990 in the United States was 27.

(*Continued*)

Table 1.1 (***Continued***) Timeline for biologic discoveries leading to biotech product development

---

1994 – The first genetically modified food, a tomato, was created for commercialization.

1995 – Dr. J. Craig Venter in the United States created the first complete genetic sequence for a free-living organism, *Haemophilus influenzae.*

1998 – First antisense molecule, Vitravene, from ISIS was approved by the FDA to treat cytomegalovirus (CMV) retinitis.

1998 – A tissue therapy of skin cells, Apligraf, was first approved by the FDA for wound healing.

1998 – Separately, James Thomson and John Gerhart were first to grow embryonic stem cells.

2000 – The human genome was elucidated completely and cloned through an international consortia and, at the same time, by the biotech company Celera.

2000 – Total approved biotech products by the end of 2000 in the United States was 109, an added 82 products over this decade.

2006 – First biosimilar biotech product, Omnitrope (recombinant human growth hormone), was approved in Europe.

2009 – The FDA approved the first biological product to be produced in a transgenic animal, a goat, i.e., the anti-thrombin protein, Atryn, by GTC Biotherapeutics.

2010 – Total approved biotech products by the end of 2010 in the United States accelerated to 221, an added 112 products over the prior decade.

2012 – The first gene therapy was approved in Europe, Glybera, from unicure for lipoprotein lipase deficiency.

2012 – Gene editing system, CRISPR-Cas, was described.

2017 – Chimeric Antigen Receptor T-cell therapy (CAR-T), Kymriah, from Novartis was approved by the FDA for acute lymphoblastic leukemia.

2018 – Ribonucleic acid inhibition (RNAi) therapy, Onpattro, from Alnylam was approved by FDA for neuropathy from hATTR.

2019 – Total approved biotech products by the end of 2019 in the United States mushroomed to 442, an additional 221 products over the prior decade in just nine years.

Description of biotechnology

Biotechnology is the use of biological systems, that is, living systems, organisms, genetic engineering, and molecular engineering, to discover new disease pathology and targets, create a biological product to address the target and ameliorate the disease, and manufacture the molecules biologically in sufficient quantities to market a pharmaceutically useful product. They are called biotech products because of the molecule's initial structural and mechanistic similarity to naturally occurring biological substances in the human body, as well as the biotechnology technologies employed in their discovery and manufacture. In discovering, creating, and manufacturing a biotech product, biotechnology faces a daunting set of challenges engaging an integrated composite of many biological and related sciences, that is, biochemistry, cell biology, embryology, engineering (seven possible types – biochemical, cellular, genetic, mechanical, molecular, process, and tissue), genetics, microbiology, molecular biology, pathology, pharmacokinetics, pharmacology, physiology, proteomics, and toxicology.

Biotechnology is now a plethora of biologic techniques and drug development technologies that permit whole new biologic discoveries and result in novel biotech products. Core basic technologies employed in biotechnology include combinatorial chemistry, high-throughput screening (HTS), bioinformatics, genomics, polymerase chain reactions (PCRs), pharmacogenomics, proteomics, and X-ray crystallography. The two cornerstone technologies in biotechnology for product creation are recombinant DNA (rDNA) technology for proteins (100 or more amino acids) and peptides (less than 100 amino acids) and, secondly, monoclonal antibodies (mabs) technology. Also, peptide products can be reproduced from nature through recombinant DNA technology, or molecularly engineered as therapeutic alternatives to large proteins. Furthermore, genetics-related technologies in the development of products employ antisense RNA (anti-RNA), aptamers, gene therapy, ribozymes, or transgenic animals. In addition, biotech virology for vaccine development involves recombinant or molecularly engineered molecules. Moreover, cells and tissues are created and formulated as therapeutic alternatives in disease management. A major group of technological advances in biological product development over the last 20 years is molecular engineering, comprised of a variety of biologic and chemical manipulations to construct new molecules, such as the following five types of manipulations:

- A biological molecule is *structurally manipulated* either with regard to its amino acid or carbohydrate content to enhance biological function or reduce toxicity, or,
- Proteins or peptides are bound together to create *fusion proteins* with combined or improved actions or an altered timeframe of actions, or,
- *Pegylation* of molecules (addition of polyethylene glycol molecules) is conducted to both alter the product's timeframe of actions and possibly create a type of shield for the molecule with less untoward immunologic adverse reactions, or,
- Multilayered *lipid molecules* are created to carry drugs or biologicals, potentially improving drug delivery with added efficacy and reduced toxicity, or,
- *Polymers* are developed incorporating a biological product or drug with other molecules to enhance a product's activity and delivery.

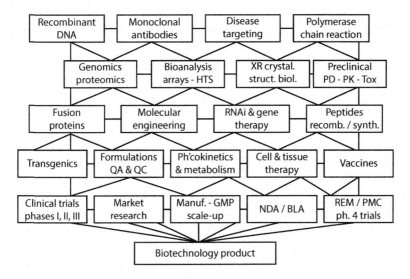

Figure 1.1 Matrix of biotechnology product development.

Abbreviations: Biol – biology, BLA – biologic license application, GMP – good manufacturing practices, HTS – high-throughput screening, NDA – new drug application, PD – pharmacodynamics, Ph – phase, PK – pharmacokinetics, PMC – post-marketing clinical trials, QA – quality assurance, QC – quality control, Recomb – recombinant, REM – risk evaluation and mitigation strategies, RNAi – ribonucleic acid inhibition, Struct. – structural, Synth – synthesis, Tox – toxicology, XR – X-ray. (With kind permission from Taylor & Francis: Evens, R.P., *Encyclopedia of Pharmaceutical Biotechnology*, 4th ed., Swarbrik, J., Ed., Taylor & Francis Group, Boca Raton, FL, 2013.)

Figure 1.1 displays many of these technologies and functional groups in a multifaceted matrix of product research and development, along with traditional drug development processes, that is, pharmacology, pharmaceutics, toxicology, and pharmacokinetics), as well as good clinical (research) practices, good manufacturing practices, and manufacturing scale-up. These technologies, processes, and functional groups culminate in new drug applications (NDAs) or biological license applications (BLAs) submitted to regulatory authorities establishing quality, efficacy, and safety for a biotech product's approval for medical use and marketing. An overarching biotech decision in product development now includes what types of products to create: recombinant proteins versus peptides, versus monoclonal antibodies, versus fusion proteins, versus pegylated molecules, versus polymers, versus cell/tissue therapy, versus vaccines, versus gene and RNA therapy, as the therapeutic choice to treat the disease. All of these technologies and processes are collectively employed by the biotechnology industry to discover, develop, and produce biotechnology products for patients. These biotech technologies collectively yield manifold scientific and health care benefits by

- Elucidating new biologic mechanisms of disease,
- Identifying naturally occurring substances or processes responsible for a biologic effect,
- Creating duplicates of the natural substances that are often found only in minute amounts in the body or nature,
- Innovating new products that enhance natural processes against disease,
- Blocking function of dysfunctional proteins or nucleic acids,
- Reducing activity of natural processes gone awry as in inflammation in arthritis,
- Offering biologic carrier molecules improving drug delivery,
- Permitting ultimately mass production of these novel products for commercialization.

About 440 biotech products have been developed and marketed from 1982 through 2019, as the result of these new "biotechnology" methods for research – development – manufacturing and are available to treat human disease. They comprise "biotherapy," a treatment armamentarium for health care providers, to be used in conjunction with drugs, devices, radiology, physical therapy, and psychotherapy for disease management.

Sciences of biotechnology

A large percentage of the biotechnology products (about 60%) are proteins and have been created through rDNA and mab technologies, which will be discussed in the next two chapters. Another product category is responsible for many new molecules and products, that is, peptides (25% of products). Biotechnology has expanded substantially in its technologies over the last 30 years to encompass a breadth of areas, many of which are listed in Figure 1.2, and will be addressed in subsequent chapters. The technologies are listed as waves of new science discoveries and tools, first to fourth, but actually overlap significantly as well, and all continuing until today.

Current biotechnology product status

From the 1980s through 2018, the 404 biotechnology-related products (discovered, developed, manufactured, and marketed) commonly have been proteins, composed of 11 uniquely different types, including blood factors, enzymes, growth factors, hormones (especially for diabetes, fertility, growth), interferons, interleukins, monoclonal antibodies (especially in cancer and immune disorders), antibody derivatives, fusion proteins,

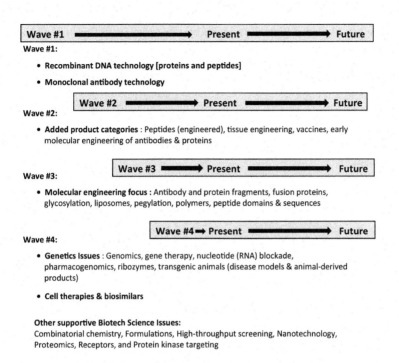

Figure 1.2 The technologies of biotechnology contributing products.

protein toxins, and transgenic-derived proteins (see Table 1.2). Peptides are found in the body and nature, which impact normal tissue function, and have commonly been employed or engineered to correct abnormal disease pathology (100+ products). The focus on proteins' development in biotechnology and disease management is predicated on their ubiquitous presence and manifold functions in human health, that is, enzymes and hormones for *tissue and cell function; signaling roles* between cells and tissues; *intracellular roles* of signaling, chaperones, degradation, differentiation, and reproduction; *immune system* involvement to maintain normal tissue function and thwart external threats to cells and tissues; and *structural employment* in all tissues to give cells and tissues their structure. Abnormal, missing, or excessive proteins are responsible for a quite large number of human diseases. Additionally, biologicals have expanded greatly beyond proteins and peptides into a very broad variety of molecules to mitigate disease, including DNA molecules (gene therapy), RNA derivatives such as RNA inhibitors, plus vaccines, tissue therapies, cell therapies, and biologic drug carriers such as liposomes or polymers.

In many diseases, biotechnology products often have been major breakthroughs offering the first treatments where nothing previously

Table 1.2 Marketed biotech products by categories (as of December 23, 2019)

Types of products	Molecules	Products	Indications[a]
Proteins/recombinant			
• Hormones[b]	10	26	31
• Enzymes[b]	31	31	23
• Growth factors	16	22	19
• Interferons[b]	10	13	14
• Interleukins[b]	4	4	5
• Blood factors[b]	28	30	9
• Fusion proteins[b]	26	29	22
Proteins toxins	5	5	15
Protein monoclonal antibodies	89	109	127
Vaccines (Recomb./Mol. Eng.)[b]	27	28	18
Peptides (Recomb./Mol. Eng.)[b]	92	112	45
Liposomes (Mol. Eng.)	11	13	13
Oligonucleotides (RNAi/Gene tx.)	12	12	11
Cell therapies	6	6	11
Tissue therapies	11	12	8
Polymers[b]	9	11	14
Others	2	2	2
Totals[c]	368	442	380

Abbreviations: Mol. Eng. – Molecular engineering, Recomb. – Recombinant, RNAi – Ribonucleic acid inhibition, tx – Therapies.

[a] Totals for the indications do not sum all the rows above because the same uses/indications were repeated for several different product types.
[b] One product may fit into more than one biotechnology category.
[c] A molecule of product may fit into more than one category above, but the total tallies do not double count and only total the individual number of molecules or products.

was effective for serious and even rare diseases, for example, imiglucerase for Gaucher's disease, human papilloma virus vaccine for cervical and vulvar cancer prevention, beta-interferon for multiple sclerosis, alteplase for acute myocardial infarction, or pavilizumab for respiratory syncytial virus (RSV) pneumonia infections. Furthermore, biotech products are highly effective in various chronic immune diseases, such as rheumatoid arthritis, ulcerative colitis, and psoriasis, providing higher patient response rates, disease control (symptoms and signs of disease), and even arrest of disease progression. Now, we have biological products in more than 14 distinct categories. Table 1.2 presents a statistical overview of the number of biotechnology molecules, biotechnology products, and number of indications. The 368 distinct molecules and 442 biotech products are used for about 380 separate indications and were created by

140 traditional biotechnology companies and the support of 20 pharma companies. Monoclonal antibodies are a very commonly created product category products (109), given their exceptionally broad applications and high specificity to diseases in all medical disciplines, especially manifold cancers and many immune diseases. Seven protein categories rank collectively high in the number of commercial products available (144 products in all), involving hormones, enzymes, blood factors, growth factors, cytokines (interleukins and interferons), fusion proteins, and toxin proteins. Peptides are a very large segment of biotech products (112 products). Vaccines are relatively common, also including recombinant forms and molecularly engineered molecules.

Biotechnology further encompasses biological products that have agricultural uses and industrial applications. Farming is being revolutionized such that genetically modified organisms in plants are being used in hundreds of millions of acres of food crops with manifold farming and public benefits; for example, greater crop yields per acre, less insecticide and herbicide use for an improved environmental impact, crop growth in stressed environments (low water and saline soil), and less cost per acre for farming. In industry, naturally occurring microorganisms are being studied and used to consume substances harmful to the environment, such as hydrocarbons (e.g., oil), mercury, and sulfuric acid. Biodegradable enzymes are used in manufacturing to replace toxic substances that previously entered the biosphere.

This book of biotechnology, in subsequent chapters, especially addresses the sciences of biotechnology and the major product categories. The sciences covered are responsible for the discovery and manufacture of biotech products including recombinant DNA technology for peptides and proteins, monoclonal antibodies, molecular engineering (peptides, monoclonal antibodies, liposome and polymers), genetic technologies with inhibitory RNA and gene therapy, and other areas (such as cell and tissue therapies). The products are covered, based on their types/categories, as well as their patient care utilization. Also, the book addresses health care applications, government engagement, and the business of biotechnology. Government engagement impacting biotechnology is reviewed, especially the regulations and laws for research, product approval, and product utilization. The companies engaged in research and commercialization are discussed, including who, what, and where, financing, alliances, mergers and acquisitions, and sales of products.

chapter two

Science of biotechnology – Recombinant DNA technology

The cornerstone of biotechnology is and has been recombinant DNA technology (rDNA) over the last 40+ years from 1970s and will continue well past 2020. The evidence for this is substantial and manifold, such as, rDNA (a) establishing the science of biotechnology as a viable means to create and produce effective and safe molecules for patient care in previously untreatable diseases; (b) being responsible for the first biotechnology product, human insulin in 1982, and fostering up to nearly 200 recombinant protein products currently approved for medical use (plus peptides, monoclonal antibodies, and some vaccines); (c) being a profitable technology for new products ($65 billion in sales in 2005, up to $260 billion in 2018) to encourage further companies and products. Proteins have been identified in the human body for their beneficial or malevolent actions and then created *ex vivo* to be used later as a therapeutic agent to interact with human physiologic systems addressing a human disease. Proteins are ubiquitous in both numbers and function in nature and the human body, see Table 2.1, suggesting many therapeutic roles.

Proteins are complex, large molecules with several key structural features required for their biologic activity. Each protein has a specific amino acid sequence from the 20 amino acids in nature, numbering usually in the hundreds to thousands, which needs to be preserved intact. Special structural features, also often known as posttranslational modifications, include multiple disulfide bridges across the amino acid chain, several special peptide domains often engendering functions to the protein, specific terminal amino acid and carboxyl species, glycosylation (carbohydrate species attached to the protein backbone), isoforms with naturally occurring protein variations in structure, and highly specific three-dimensional (3D) configurations necessary for function, especially receptor interactions. Figure 2.1 presents the structural elements as described above for the protein alteplase, a thrombolytic enzyme indicated to avoid blood clots in acute myocardial infarction. The protein structure for alteplase includes 527 amino acids in its chain (pictured as a sequence of tiny circles for each amino acid in the chain, plus the total noted by a hexagon in the figure), three individual glycosylation sites at amino acids

Table 2.1 Roles of proteins in human anatomy/physiology/disease

Structure:
  - Tissues: muscle, bone, epithelium, endothelium, skin, nerves, hair, blood, cells
  - Organs: heart, kidney, liver, pancreas, glands, intestines, eyes, pituitary, brain, spinal cord

Communication (intercellular):
  - Hormones, stimulating factors, interleukins, interferons, receptors

Functional:
  - Catalysis (enzymes), blood clotting, glucose processing, lipid processing, protein processing

Immune defense/reactions:
  - Three categories of interferons and 30+ interleukins
  - Immune processes of identification, recognition, tagging, adhesion, removal
  - Involving many cell types; neutrophils, macrophages, T-lymphocytes, B-lymphocytes, natural killer cells, eosinophils

Cell function (intracellular):
  - Signaling, receptors, chaperones, degradation, differentiation, reproduction, energy production, apoptosis, necrosis

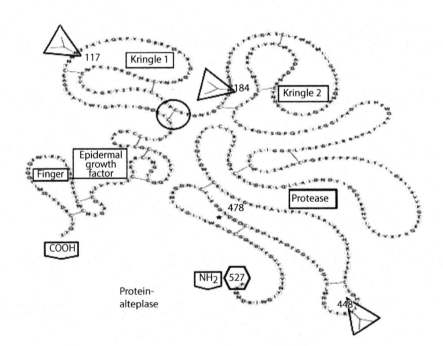

Figure 2.1 Protein structure, alteplase. (Reprinted with permission from Amgen Inc.)

117, 184, and 448 (triangles in the figure), disulfide bridges (#15 of them) (one example noted by a circle), end amidation and end carboxylation (pentagons), and 5 structural peptide domains (finger, kringle 1 and 2, protease, and epidermal growth factor – rectangles in the figure), which impart physiologic function and structure to the protein.

Yet another structural feature for proteins is the highly specific and complex folding process of all proteins into their 3D structure required for normal functioning in their interaction with cell receptors and other proteins and peptides. As an example, Figure 2.2 presents the protein filgrastim in its 3D structural form, with the folding of the amino acid chain into four alpha helixes and other structures.

Figure 2.2 Protein three-dimensional structure. Filgrastim (granulocyte colony stimulating factor). (Reprinted with permission from Amgen Inc.)

Table 2.2 Tools of recombinant technology

1. Gene isolation	2. Cloning & expression	3. Protein production
Gene (DNA)	Plasmid (DNA)	• Clone
Reverse transcriptase enzyme	Restriction endonuclease enzyme	• Growth media
Polymerase enzyme	Ligase enzyme	• Fermenter
DNA probe	Promoter/enhancer DNA	• Roller bottles
	Linker DNA	• Centrifuge
	Host cell	• Chromatographer
		• Buffer/Stabilizer

Source: Reprinted with permission from Amgen Inc.

The basic notion in rDNA technology is the manufacture of human proteins from human genes in nonhuman living systems, a form of genetic engineering. A variety of tools necessary for rDNA technology and genetic engineering to manufacture proteins are summarized above in Table 2.2, displayed in three categories: (1) gene isolation, (2) cloning and expression, and (3) protein production. Symbols represent the types of tools; *keys* are for genetic enzymes that manipulate various DNA components; a *double helix* is for DNA material directing transfer of genetic code; *circular DNA* is for plasmids that carry DNA between cells and for manipulation of genetic codes; and a *cell* for a host cell that is the entity in which rDNA is performed and actual manufacturing occurs. These tools will be discussed first to give context and description of what they are and how are they used, prior to discussing the steps of the rDNA process and how they are employed.

For background and context, several tenets of genetics underpin rDNA technology. First, chromosomes (46 in humans) in the nuclei of cells contain all the genes (about 30,000 in humans) to be found in cells. Genes are comprised of DNA. See Figure 2.3. DNA is configured in two matching strands of DNA in a double helix in a gene, including matched pairs (about 3 billion, A-T & G-C) of four nucleotides, A – adenine and G – guanine (purines), and T – thymidine and C – cytosine (pyrimidines).

For the second tenet, the central dogma of biology and rDNA technology is that DNA is, first, transformed into RNA in the nucleus through the process of *"transcription"*; second, followed by RNA being transformed into proteins in the cell cytoplasm through the process of *"translation"*; followed by, third, cytoplasmic *"posttranslational modifications"*

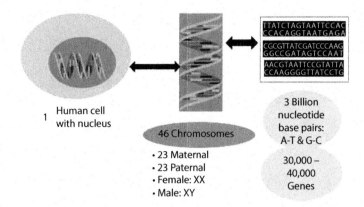

Figure 2.3 Chromosomes, genes, DNA. (Reprinted with permission from Amgen Inc.)

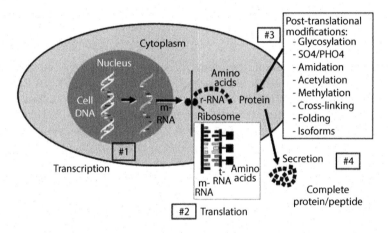

Figure 2.4 Central dogma of biology and rDNA technology.

Abbreviations: m-RNA – messenger RNA, PHO4 – phosphate, rDNA – recombinant DNA, r-RNA – ribosomal RNA, SO4 – sulfate, t-RNA – transfer RNA. (Reprinted with permission from Amgen Inc.)

of the protein turning it into a complete protein, which, fourth, ends in *"secretion"* from the cell, as displayed in Figure 2.4. The translation process involves three types of RNA, that is, messenger RNA (m-RNA) carrying the genetic code from the nucleus to the cytoplasm, ribosomal RNA (r-RNA) in the cytoplasm creating the structure in which the genetic code in the m-RNA can be read to create proteins and also involving transfer RNA (t-RNA), which brings the amino acids to the ribosome to match with the m-RNA code. t-RNA involves nucleotide triplets for each amino acid that will match up with the nucleotide code of the m-RNA, bringing the correct peptide for the protein chain to be extended.

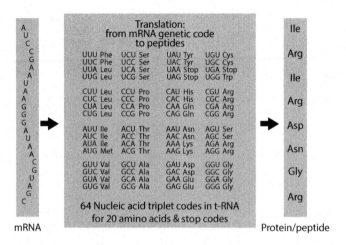

Figure 2.5 Triplet genetic codes in t-RNA.

Abbreviations: m-RNA – messenger RNA, t-RNA – transfer RNA. (Reprinted with permission from Amgen Inc.)

The interpretation of the genetic code from m-RNA to proteins involves triplets of nucleotides, of which 64 triplets for the t-RNA exist for the 20 amino acids, including a start code triplet to initiate translation and a stop code triplet ending translation, which are depicted in Figure 2.5.

Another tenet in genetics of rDNA technology and genes involves the translation process, wherein the genes in chromosomes contain many translating DNA sequences called exons and also interceding sequences called introns that are not translated, which is shown in Figure 2.6. As DNA is translated and transformed into m-RNA, only the exons are incorporated and united through action of the spliceosome into the coding

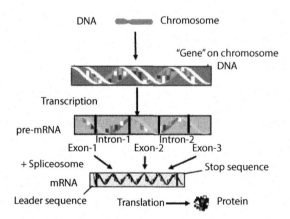

Figure 2.6 Exons and introns in genes. (Reprinted with permission from Amgen Inc.)

m-RNA strand. Also, a nucleotide alteration occurs during translation, in that the nucleotide uracil in m-RNA is substituted for thymidine in DNA. Furthermore, the gene will contain a leader m-RNA sequence and a stop m-RNA sequence for turning on and off the translation process, respectively, in creating proteins from genes.

The gene found to produce the desired protein from any method is then multiplied exponentially using the enzyme DNA polymerase in order to have a sufficient number of genes to continue with rDNA technology. As demonstrated in Figure 2.7, three steps are involved in the manifold duplication of genes; first, the DNA double helix is denatured with heat at 90°C, separating the DNA double helix into single chains. Second, primers are annealed to the target DNA segment (gene) for both chains. Third, the target gene with primer is extended with nucleotides (adenine, guanine thymidine, cytosine) to yield the gene, under the action of DNA polymerase enzyme to foster the process. Repeating this three-step process 35 times creates about one million copies of the gene.

In rDNA technology, a bacterial plasmid is employed that is a circular piece of DNA that normally can be transferred between cells (therefore, a DNA carrier), depicted in Figure 2.8. Plasmids in nature carry genes between bacteria; for example, antibiotic resistance genes from one cell to another cell gives the new cell a new property of resistance to an antibiotic. Plasmids will accept the insertion of a human gene, will allow the human gene to be turned on and function, and must ensure gene production of proteins (expression) over time and without alteration. Plasmids are further manipulated to maximize their function with the addition of a viral DNA promoter sequence and a viral DNA enhancer sequence, which amplify the cells' genetic functioning. Other DNA pieces are included, such as an operator DNA sequence

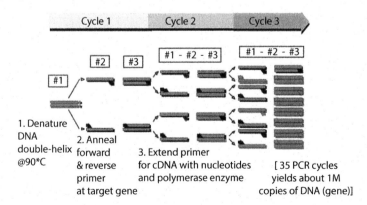

Figure 2.7 Gene reproduction – polymerase chain reaction.

Abbreviations: cDNA – complementary DNA, PCR – polymerase chain reaction. (Reprinted with permission from Amgen Inc.)

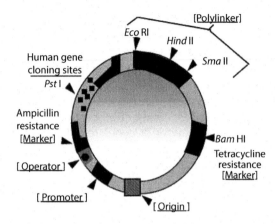

Figure 2.8 Bacterial plasmids (DNA carriers). (Reprinted with permission from Amgen Inc.)

improving plasmid function. A linker piece of DNA may be needed to close the plasmid DNA circle. Some bacterial plasmids will incorporate the human gene, and some do not, such that you need to ferret out the plasmids containing human genes from the mixture of plasmids. By also adding antibiotic resistance genes into the plasmid along with the human gene, you then can add antibiotics, such as ampicillin and/or tetracycline, to the plasmid mixtures in cell culture, which will destroy all plasmids without the resistance markers, resulting in only the desired plasmids with the human gene remaining. Plasmids are incorporated into host cells, a process called *"transfection."* Plasmids can continue to function over time in culture. The ideal set (#12) of the many functional properties for plasmids are listed in Table 2.3.

Table 2.3 Summary of properties of plasmids

- A circular piece of DNA, active in cells
- Source – bacteria
- Accept human genes into plasmid structure
- Vectors (carriers) for human genes
- Accept viral genes (enhancers/promoters) into plasmid structure
- Turn on & turn off the target human gene
- Replicate in cells
- Not change expression of human protein
- Insert (transfer) into various host cells (process of transfection), such as bacteria, mammalian cells, plant cells
- Function over an extended timeframe of cell culture (producing human proteins)
- Off-target effects minimal (e.g., interferons production minimal)
- Intellectual property (patentable step in manufacturing products)

The circular plasmid DNA must be cut open to accept the human DNA (gene) using unique bacterial enzymes (restriction endonucleases). Each endonuclease enzyme is highly specific to a certain nucleic acid sequence, creating a very specific cut, that is, an opening in the DNA plasmid structure appropriate for a specific gene's incorporation and permitting efficient recombination. DNA materials will recombine naturally with the human gene sequence inserted into the circular plasmid sequence. A DNA ligase enzyme is employed to enhance the DNA recombination process. Figure 2.9 displays such restriction endonuclease enzymes found in nature in specific bacteria as noted. The very high specificity to an individual DNA sequence of nucleotides for endonucleases is shown in Figure 2.9.

Host cells in rDNA are employed to incorporate the plasmids containing the human gene and allow plasmid function and protein production. Host cells need to possess a large set (#10) of demanding characteristics to be used feasibly and cost-effectively in rDNA manufacturing processes as follows: (1) a short reproductive life cycle, allowing multiple cycles over time favorably lengthening the duration and quantity of protein production; (2) long-term viability in an *in vitro* setting; (3) the ability to accept bacterial plasmids; (4) the ability to allow the plasmids to function normally within the cells; (5) a substantial productive capacity (yield) for proteins; (6) the ability to produce the human protein consistently and without its alteration; (7) the ability to possibly perform glycosylation of the protein, if needed for its activity; (8) a relative ease of handling and manufacturing in the later scale-up process; (9) as low as possible cost in manufacturing; and (10) patentability to protect the intellectual property. The host cells can be bacteria, usually *Escherichia coli*; yeast cells, most

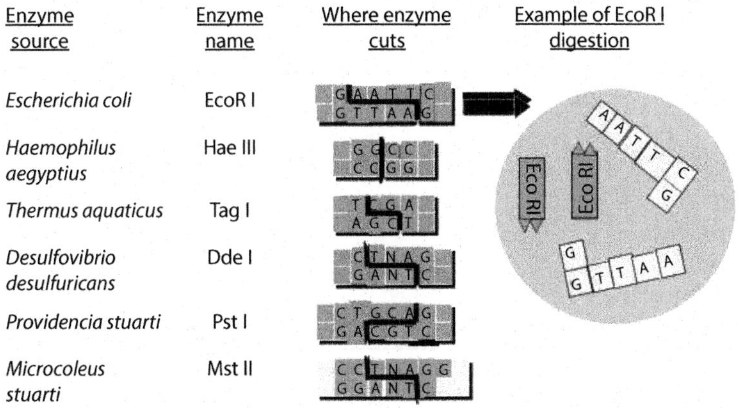

Enzyme source	Enzyme name	Where enzyme cuts	Example of EcoR I digestion
Escherichia coli	EcoR I	G A A T T C / G T T A A G	
Haemophilus aegyptius	Hae III	G G C C / C C G G	
Thermus aquaticus	Taq I	T C G A / A G C T	
Desulfovibrio desulfuricans	Dde I	C T N A G / G A N T C	
Providencia stuarti	Pst I	C T G C A G / G A C G T C	
Microcoleus stuarti	Mst II	C C T N A G G / G G A N T C C	

Figure 2.9 Restriction endonucleases – bacterial sources & DNA sequence specificity. (Reprinted with permission from Amgen Inc.)

often *Saccharomyces cerevisiae*; mammalian cells, usually Chinese hamster ovary cells (CHO) or baby kidney hamster cells (BKH); or genetically modified plant cells, for example, carrot or tobacco. The choice of the best host cell depends on the protein and the manufacturing process desires of the company. These unique newly created host cells and their offspring, created in the laboratory, are called the "master cell working bank." See Table 2.4.

Certain proteins such as epoetin alpha, must be fully glycosylated (carbohydrate structures attached to proteins) to be physiologically active, and only mammalian host cells (CHO and BKH cells) can perform the complete glycosylation process for many proteins. Although all proteins are naturally glycosylated in the human body, many proteins possess full activity without glycosylation, such as filgrastim [granulocyte-colony stimulating factor (G-CSF)]. Such proteins in their rDNA processes can employ non-mammalian host cells such as

Table 2.4 Host cell properties

Cell types ➡					
Property ▼	CHO	GMO plants	Yeast	BHK	Bacteria cells
Productivity	+++	++	++	+	+++
Secretion	Yes	No	Yes/no	Yes	No
Protein folding	Correct	Some	Some	Correct	Misfold
Nutritional needs	Complex	Simple	Simple	Complex	Simple
Growth rate	Slow	Medium	Fast	Slow	Fast
Glycosylation	Yes	Yes, but	Yes?	Yes	No
Product quality	++	++	+	+	+
Impurities	++	+ ?	+	++	+/−
Ease of manufacture	+++	++	++	++	++++
Economics	++	+++	++	+	++++
Time to clinic	+	++?	+	+	++
Intellect. property	+	++?	+	+	+
Regulatory	+++	++?	++	+++	+++

Source: Adapted from Kiss, R.D., *Gen. Eng. News*, 24, 1, 2004; Pallaoathu, M.K., *Gen. Eng. News*, 26, 552, 2006. With permission.

Abbreviations: BHK – baby hamster kidney, CHO – Chinese hamster ovary, GMO – genetically modified organism, intellect – intellectual.

Note: The number of [+] signs indicate the relative increasing significance of the feature.

bacteria or fungal or plant cells. These types of host cells are easier to reproduce and control. Patents are an important further advantage in rDNA manufacturing for biopharma companies from a legal perspective. In the creation of biotech products, patents can be obtained for new and unique host cells or plasmids or other technical improvisations. These additional patents in manufacturing of a biotech product offer added legal protections for a molecule that is approved for use, preventing copycats reaching the marketplace until the added patents lapse.

Another tool in rDNA manufacturing is the culture media in which the host cells must live, function, and produce its proteins. Fairly standard contents for culture media include serum, sucrose or trehalose, minerals (calcium, potassium, sodium); plus, the pH, tonicity, and temperature must be controlled. Also, to enhance host cell viability and function, the media needs to be exchanged during the manufacturing process with new media to feed the cells. Second, the media needs to be removed, that is, harvested periodically, as it contains the proteins produced by host cells, including the human protein target of the manufacturing process. Mammalian cells secrete the proteins into the media, allowing media collection, but bacteria and fungal species as host cells maintain the newly produced proteins within the cells, usually in vacuoles, requiring harvesting of the cells and shearing them open to obtain the proteins. Media collection also needs to eliminate the other compounds and impurities normally secreted by the host cells The nine desirable properties for the culture media are itemized in Table 2.5.

These genetic and protein tools are employed in the following steps that comprise rDNA technology; (1) *protein* identification-isolation-characterization, (2) *gene* (specific for the protein) isolation-reproduction, (3) *cloning* of genetic material and *expression* of proteins, (4) *manufacturing* for commercialization (*scale-up* processes), and (5) *quality assurance* for the protein, protein product, and process integrity.

Table 2.5 Culture media in rDNA – properties

- Host cell viability over long timeframes (weeks to months)
- Host cell productivity for protein production
- Protein posttranslational modification permitted
- Protein stability without degradation or aggregation
- Contamination low
- Impurities low (host cell & serum proteins, residual DNA)
- Ease of handling
- Reasonable cost for production
- Disposable into nature without harm to the environment

Abbreviation: rDNA – recombinant DNA.

Step 1 of the rDNA process involves finding a protein or peptide responsible for some biological effect in the human body that has therapeutic potential. The protein needs to be isolated from its normal milieu, usually a body fluid or cell. The structure of the protein and its function are determined, including all the structural features noted above. Many proteins may be responsible for the creation or function of cells, such that a challenge exists to find the primary protein or proteins responsible. Figure 2.10 shows the complexity of hematopoietic cell production from 1 pluripotent stem cell evolving into 11 different cells, each with distinct functions, and directed by different sets of proteins at each stage of cell evolution. Only the proteins engaged in red blood cell formation are itemized at each step of cell evolution in this figure [stem cell factor (SCF), interleukin-3, granulocyte-macrophage (GM)-CSF, epoetin alpha]. Please note that the many other proteins involved in producing all the other hematopoietic cells are not listed in the diagram for simplicity of the diagram. These proteins are called collectively Colony stimulating factors (CSFs) and include such as epoetin alpha in red blood cell production. CSFs can be lineage specific, such as epoetin alpha only impacting red blood cell function, or CSFs

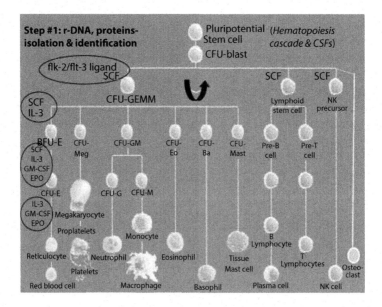

Figure 2.10 Step 1: rDNA, proteins identification and isolation.

Abbreviations: Ba – basophil, CSF – colony stimulating factor, CFU – colony forming unit, Eo – eosinophil, EPO – epoetin alpha, IL-3 – interleukin-3, GEMM – granulocyte-erythrocyte-macrophage-megakaryocyte, GM – granulocyte-macrophage, NK – natural killer cell, rDNA – recombinant DNA, SCF – stem cell factor. (Reprinted with permission from Amgen Inc.)

can be nonspecific, such as SCF involved in the formation of all cell lineages and at the early stages of cell evolution.

Step 2 requires the isolation of the target human gene responsible for making the target protein, which entails one of the three gene processing mechanisms: (1) reverse engineering from the protein sequence, using genetic codes to the DNA sequence (gene); (2) reverse engineering from mRNA to complementary DNA; and (3) gene fishing in the human genome. See Figure 2.11.

For the first method to find the human gene, we often will know the protein's full amino acid sequence, and the 64 nucleic acid triplets that code for the 20 amino acids. Thus, we can construct many combinations of the triplet codes that genetically represent the possible target gene for the target protein. These genetic constructs are genes, one of which will be identified through pharmacodynamic screening as the correct gene with the capacity to produce the target protein. Second, we may be able to find the human cell that produces our target protein. In this cell, we can ferret out the mRNA that is responsible for producing the target protein through the process of translation. The viral enzyme, reverse transcriptase, is capable of creating in reverse the target complementary DNA, or gene, from this specific mRNA. This method was used to find the gene for insulin, which led to the marketed product of recombinant human insulin. Third, the human gene could be fished out of the human genome using nucleic acid probes, which is a complex, daunting task. In this method, we identify several peptides in the amino acid sequence of the protein. Using the aforementioned triplet codes for the amino acids in the peptide subunits of the protein, we build a specific nucleic acid combination (probe) for each peptide. Then, we break the chromosomes into

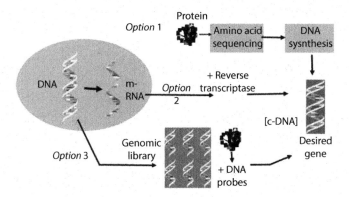

Figure 2.11 Step 2: rDNA gene isolation.

Abbreviations: cDNA – complementary DNA, m-RNA – messenger RNA, rDNA – recombinant DNA. (Reprinted with permission from Amgen Inc.)

thousands of pieces of DNA. Through many serial experiments, we try to match the first nucleic acid probe to the DNA mixture, which creates a subset of matching DNA pieces (hundreds or thousands). In another series of experiments, a second, different nucleic acid probe for a different peptide is matched against this DNA subset for further matches. Matches do occur, resulting in a series of possible genes. Each gene must be evaluated by genetic analyses to ensure production of the correct protein, which is then tested to be sure that it has the structural features and pharmacological properties in test animals of the targeted, naturally occurring protein. The gene for the protein epoetin alfa, used for anemia, was discovered by this laborious method.

Step 3 in rDNA technology involves *cloning* of the gene and *expression* of the protein by the gene in host cells. Cloning is the reproduction and multiplication of the new nonhuman cell containing a human gene in the plasmid in the cell; a group of the new cells that is reproduced is called a "clone." Expression is the production of the target human protein by a nonhuman host cell containing the human gene. These processes require a vector for the DNA (genes), that is, plasmids, as discussed earlier, so that the gene can be carried into a host cell.

Within Step 3 of rDNA, we have six further steps; plasmid selection, gene incorporation into the plasmid, enhancer/promotor DNA incorporation into plasmids, vector transfer to host cells, cloning of cells, and expression of the protein, as delineated in Figure 2.12. Plasmid selection and manipulation for optimal functionality is addressed previously with the discussion for Figure 2.8. Second, endonuclease enzymes cut open the plasmid at specific locations for inclusion of the

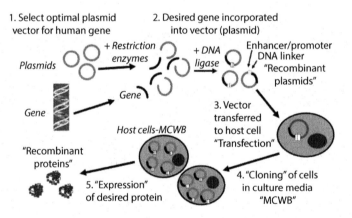

Figure 2.12 Step 3: rDNA cloning and expression with plasmids and host cells.

Abbreviations: MCWB – master cell working bank, rDNA – recombinant DNA. (Reprinted with permission from Amgen Inc.)

human gene. The "sticky" ends of the opened bacterial plasmid and the human gene permit recombination of the DNA, under the influence of a ligase enzyme, resulting in a recombinant DNA molecule containing a human gene inserted into a bacterial plasmid. Third, the rDNA molecule (plasmid plus human gene) is inserted into a host cell (transfection), which serves as the manufacturing site, producing all of host cell's routine proteins; plus, the cell manufactures the human protein from the human gene that it carries. Fourth, biotechnologists then perform "cloning" of the newly created cells (a cell containing the plasmid and human gene), that is, reproduction of the host cells, which also multiplies the number of plasmids in each cell. Hence, the master cell working bank of genetically modified host cells is created. Fifth, the host cells are cultured in fermenters, and they produce the desired human protein, a process called "expression." The desired human protein must be extracted from the protein milieu. For example, high-pressure liquid chromatography is one method employed for protein separation. Then, the protein undergoes extensive testing to ensure that, in actuality, it is the desired protein with the necessary properties discussed previously.

Step 4 in rDNA technology requires cell-line development and is the scale-up process for manufacturing, comprised of four phases: Inoculum, Fermentation (plasmids and genes), Purification, and Formulation, which are demonstrated in Figure 2.13.

The first phase, "*Inoculum*," involves removal a sample of cells from the master cell working bank, called daughter cells or a clone. These daughter host cells are placed in a series of increasing larger flasks, serving as small fermenters, with culture media to permit their growth and reproduction, plus ensuring their optimal functionality. Mammalian host cell fermentation is shown in this example, and epoetin alpha protein is being manufactured.

The second phase in this mammalian rDNA is "*fermentation*," wherein the host cells are inoculated into larger vessels called fermenters. In the second phase, fermentation, the use of relatively smaller fermenters (1–2 L) are employed for epoetin alpha production, called roller bottles, because they lie on their side and roll, bathing the host cells in culture media, enhancing cell productivity and desirably limiting the volume of culture media required. The host cells secrete the proteins into the culture media. A very carefully created environment in the fermenters includes the air mixture in the bottles and appropriate culture media contents, which are designed for the type of host cell, type of fermenter, and type of protein being produced. Media is shifted, with new media added to maintain optimal host cell viability and function. Media is also withdrawn (harvested) at intervals to obtain the proteins in the media. The collected media, which is full of the various proteins

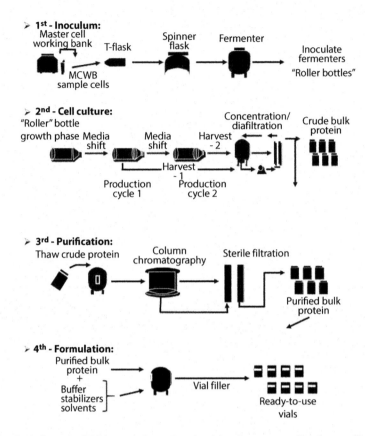

> **1ˢᵗ - Inoculum:**
> Master cell
> working bank T-flask   Spinner flask   Fermenter
>
> MCWB
> sample cells   Inoculate fermenters
>   "Roller bottles"

> **2ⁿᵈ - Cell culture:**
> "Roller" bottle   Concentration/   Crude bulk
> growth phase Media   Media   Harvest   diafiltration   protein
>   shift   shift   - 2
>
>   Production   Production
>   cycle 1   cycle 2
>   Harvest - 1

> **3ʳᵈ - Purification:**
> Thaw crude protein   Column   Sterile filtration
>   chromatography
>
>   Purified bulk
>   protein

> **4ᵗʰ - Formulation:**
> Purified bulk
>   protein
>   +
> Buffer   Vial filler
> stabilizers
> solvents   Ready-to-use
>   vials

Figure 2.13 Step 4: rDNA manufacturing (scale-up) – mammalian host cells.
Abbreviations: MCWB – master cell working bank, rDNA – recombinant DNA. (Reprinted with permission from Amgen Inc.)

being produced by the host cells, are filtered, concentrated, and frozen to result in "crude bulk protein," which also contains the desired human protein.

Phase 3 is the *"purification"* process to separate out the desired human protein from the crude bulk milieu by liquid chromatography and filtration. We now have created pure human target protein. Phase 4 is the *"formulation"* process in which the bulk protein is mixed with solvents, stabilizers, and buffers to solubilize the protein and adjust the pH of the solution, all of which attempt to create a final product with a tolerable and stable form for administration, without degradation or alteration of the active ingredient (the protein), and with a maximum shelf life as well. About 30 potential, not uncommon problems in rDNA manufacturing of proteins all need to be avoided, as listed in Table 2.6.

Table 2.6 Biotechnology product – potentials problems in rDNA manufacturing

Physical changes of the protein:
- Clumping/aggregation
- Precipitation
- Protein inclusions

Chemical changes in the protein:
- Reduction
- Oxidation
- Deamidation

Structural changes:
- Amino acid mutations, deletions/additions
- Conjugation
- Cross-linkage or un-linkage (disulfide bridges)
- Glycosylation or deglycosylation
- Misfolding/unfolding of protein
- Proteolysis
- Terminal amino acid variations

Formulation:
- Physical: color, particulates
- Chemical: concentration, pH, specific gravity, tonicity
- Shelf life
- Contamination:
  - Host cell materials and proteins
  - Media
  - Bacteria, fungi, virus
  - Oncogenes

Abbreviation: rDNA – recombinant DNA.

Step 5 in rDNA manufacturing is the extensive and sophisticated "quality control and quality assurance" analyses required. These complex molecules, proteins, with their manifold characteristics noted previously, along with the complex multifaceted elegant rDNA manufacturing processes just described, necessitate many complex, sophisticated, and novel analyses with validations to ensure the quality of the final product for safe and effective handling and use for patients. The quality control and quality assurance testing can involve about 70 possible tests in four components: (1) plasmids and host cell cells, (2) bulk protein product, (3) process validation, and (4) final product batches, which are outlined in Figure 2.14 and Table 2.7.

The five steps in recombinant DNA manufacturing of proteins and peptides in this chapter are summarized further below in tabular form in Table 2.8.

The end product of biotech manufacturing employing recombinant DNA technology is a recombinant human protein or peptide. Manufacturing in rDNA technology truly can be stated to be an elegant,

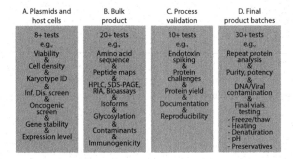

Figure **2.14** Step 5: Examples of quality control & quality assurance testing in rDNA manufacturing.

Abbreviations: HPLC – high-pressure liquid chromatography, ID – identification, Inf. Dis. – infectious disease, rDNA – recombinant DNA, RIA – radioimmunoassay, SDS-PAGE – sodium dodecyl sulfate–polyacrylamide gel electrophoresis. (Reprinted with permission from Amgen Inc.)

Table **2.7** Sample quality control tests in rDNA manufacturing

Genetic material:
- Karyotypic analysis
- Oncogene screening
- Gene stability
- Infection DNA screens
- Viral contamination

Bulk protein tests:
- Features: amino acid sequence, peptide maps, glycosylation, folding, isoforms
- Tests: high-pressure liquid chromatography and/or radioimmunoassay, Western blot chromatography, bioassay (PD – Pharmacodynamics)

Process validation:
- Protein yield
- Protein challenge
- Media quality
- Endotoxin spiking
- Reproducibility

Final product:
- Protein analyses as above for bulk protein repeated
- DNA/protein/viral/bacterial contamination repeated
- Stability tests over time
- Freeze-thaw and high temperature stress testing
- Formulation: physical examination
- Formulation: physio-chemical traits, e.g., tonicity, pH, etc.
- Glass vs. plastic bottling & stopper material impacts on protein

Abbreviation: rDNA – recombinant DNA.

Table 2.8 Recombinant DNA technology

Step 1. Protein workup:
- Protein identification
- Protein isolation and sequencing
- Biological property description (pharmacodynamics and pharmacokinetics)
- Protein structure workup – amino acid sequencing and mapping, disulfide bridging, peptide domains, glycosylation, terminal amino acids and carboxyl groups, peptide folding, isoforms

Step 2. Gene isolation (three alternative methods):
- For amino acid sequence in a protein, nucleic acid triplet combinations (#64) for each amino acid strung together to yield gene sequences
- Isolation of mRNA for protein from its source cell and using viral reverse transcriptase enzyme to transform RNA into complementary DNA (gene)
- Several DNA probes, representing a few peptides of the protein, used for matching with DNA fragments in the human genome library and fishing out the target gene

Step 3. Cloning and expression:
- Plasmids opened with restriction endonuclease enzymes
- Gene (human) insertion into bacterial plasmids (DNA vectors)
- Plasmid manipulation for expression enhancement with insertion of viral DNA promoters and enhancers
- Plasmid incorporation into host cells
- Growth media augmentation and host cell reproduction (cloning)
- Host cell production of proteins (expression)
- Separation of human protein from the protein milieu produced by the host cells
- Master cell working bank (MCWB) of cells possessing a human gene created in a nonhuman cell for production of the target protein

Step 4. Manufacturing scale-up:
- Inoculum stage, sample of MCWB grown to initiate cell activity and cell testing
- Fermentation or cell culture stage (larger volumes) for protein production with bulk protein harvesting
- Protein isolation and purification stage (chromatography and centrifugation)
- Formulation stage with protein plus diluents, stabilizers, pH adjustment and vial filling for a final commercial product

Step 5. Quality assurance with extensive lab testing:
- Genetic testing of system
- Bulk product testing (proteins)
- Process validation
- Final protein assessment
- Product testing for protein and formulation in final vial

---

elaborate, and scientific project, employing potentially highly variable living, yet highly controlled, systems, for each a lot of proteins are created while utilizing extensive process and product testing to ensure reliability and safety of the products.

chapter three

Science of biotechnology – Monoclonal antibody technology

Monoclonal antibodies (mabs) are complex large proteins produced by a white blood cell (WBC) called a "plasma cell" in the lymph system as part our immune defense system. A foreign or abnormal protein or substance, or infectious molecule, or a cancer can be an antigen (Ag) that stimulates our normal immune system. A certain type of WBC B-cell type in the lymph system is stimulated to act against the target antigen and functions as antigen presenting cells (APCs). The APCs will endocytose the antigen and internally bind the antigen to its major histocompatibility complex II molecules (MHC II). APC cells include, for example, dendritic cells and phagocytes. APCs carry the antigens to CD4+ T-lymphocyte cells that then bind to the MHC II-Ag complex, and in followup activate B-cells to differentiate into plasma cells in the lymph nodes or spleen. Plasma cells produce highly specific monoclonal antibodies that target only the specific target antigen and eliminate the offending foreign substance or cancer. All mabs in humans have a highly uniform basic structure comprised of two matched pairs of proteins; in each half of the mab, we have four heavy chains and two light chains, which are further defined as constant (four) and variable (two) regions, forming a "Y" configuration, see Figure 3.1.

One end of the molecule, the four variable regions, bind to the target antigen with their terminal complementarity determining regions (CDRs), whereas the other end of the molecule, the constant region with the heavy chains at the terminal Fc area, can bind to WBCs and initiate an immune reaction. The 9 structural regions and the 16 structural domains are labeled in Figure 3.1. (1) Variable regions are four in number at one end of the mab molecule; (2) the variable complement determining regions (CDRs), which bind with very high specificity and affinity to the target antigens; (3) the four pairs of constant regions (C_H1, C_H2, C_H3, C_L); (4) the cell binding sites at the terminal end of constant regions (F_c at C_H3), which upon binding can initiate immune cascade reactions such as antibody-dependent cell cytotoxicity (ADCC), or complement-dependent cell cytotoxicity (CDCC), or phagocyte-dependent cell cytotoxicity (PDCC); (5) light chains (V_L & V_H) with 212 residues on each half of the molecule; (6) heavy chains (V_H, C_H1, C_H2, C_H3) with 450 residues on each half of the molecule; (7) a hinge region

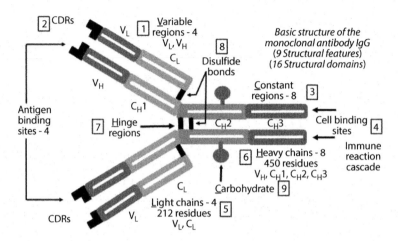

Figure 3.1 Monoclonal antibody structure.

Abbreviations: Ab – antibody, CDR – complementarity determining regions, C_H1 – constant heavy region 1, C_H2 – constant heavy region 2, C_H3 – constant heavy region 3, C_L – constant light region, IgG – immunoglobulin type G, V_H – variable heavy region, V_L – variable light region. (Reprinted with permission from Amgen Inc.)

that binds the two mirror image sections of the mab together with disulfide linkages; (8) disulfide bridges for the connection of the two constant chains (one heavy segment and one light segment) with the two variable chains together in each half of the mab structure; and (9) carbohydrate species found on the C_H2 regions of the mab, one on each half of the molecule.

Plasma cells can be short-acting (short-lived), producing IgM mabs, or longer-acting, producing IgA or IgA mabs. Mabs can be one of five isotypes, IgG (immunoglobulin A), IgD, IgE, IgG, and IgM, which vary in heavy chain structure and functions, as follows; IgA – delta heavy chains, found in mucous membranes especially in respiratory and gastrointestinal systems; IgG – gamma heavy chains, found in all body fluids (most common Ig); IgM – mu heavy chains, found in lymph and blood; IgE – epsilon heavy chains, associated with allergic reactions; IgD – delta heavy chains, with roles not well understood. In normal mab production in human body, we should further note that four genes collectively are responsible for transcription and translation into any one mab protein.

The nomenclature for the generic names of mabs is also highly structured, as displayed in Table 3.1. Four components (name fragments) are used for every mab name, all in the same order: first section, the dedicated unique name fragment for the specific mab; second, the usage area abbreviation; third, the abbreviation for the animal of origin/type of mab, and fourth, always the fragment "mab." Two representative mabs are presented in Table 3.1.

Table 3.1 Monoclonal antibody nomenclature

Examples: Tras-tu-zu-mab (Herceptin®) or Ada-li-mu-mab (Humira®)

| 1 | 2 | 3 | 4 | | 1 | 2 | 3 | 4 |

Section 2 – usage: (13)
- Cancer: tu, ta
- Immune: l(i)
- Circulatory: c(i), cu
- Infection virus: v(i)
- Bone: os, so
- Interleukin: k(i)
- Nervous syst.: n(e), nu
- Endocrine: de
- Virus: vi

Section 3 – Ab types: (9)
- Murine: mo
- Chimeric: xi 70% murine/30% human
- Humanized: zu 10% murine/90% human
- Human: mu 100% human
- Hamster: me
- Primate: mi
- Rat: ma
- Rat/Mouse: axo
- Hybrid: xizu

Abbreviation: Ab – antibody.

These mabs have the special characteristic in our immune defense of being highly specific proteins that attack a single specific antigen that is contained in a foreign or abnormal substance or cell or cancer. In concert, mabs are produced by a highly specific plasma cell, wherein one specific type of plasma cell produces a single type of mab against a single antigenic foreign material. The currently available, as of December 31, 2019, 107 mab molecules target 51 different antigens, as listed in Table 3.2. The names of the mab molecules are truncated without the mab suffix to accommodate fitting in the table. Also, the therapeutic areas (TAs) are noted for the targets. Cancer and immune diseases predominate as the usage areas for targets and mab molecules since many specific protein targets exist on cancer cells and in immune diseases. Some targets have multiple different mabs marketed to attack the antigens and ameliorate their diseases.

Additional traits of mabs are found in Table 3.3, with tabulations of six antibody types (types of mab constructs), isotypes, recombinant host cells, uses by medical disciplines, and other structural forms. Host cells for the manufacture of mabs include six cell types with Chinese hamster ovary (CHO) cells predominating (49 molecules). The uses for mabs are extensive in number (97) and broad in scope involving nearly all medical disciplines. The most common mab construct is "zu" for "humanized" mab molecules (46 out of 86), which are 90% human, 10% murine. The other traits include a variety of cell engineered molecules, which will be discussed in the science chapter for molecular engineered molecules.

Table 3.2 Antigens (targets) for monoclonal antibodies (through 12.31.2019)

Antigen	TA	Mab	Antigen	TA	Mab	Antigen	TA	Mab
BLys B.Anthracis Rh	ID	Belimu-Obiltoxaxi-, Raxibacu-	EGFR	ON	Cetuxi-, Panitumu-, Necitumu-	Kallekr. Nectin-4	IMON	Lanadelu-Enfortu-
C5	IM	Eculizu-, Ravulizu-	FGF23	Gen	Burosu-	p-Selectin	H	Crinzanlizu-
CCR4	ID	~~Antigen~~ Mogamulizu-	F9a/F10	H	Emicizu-	PDGFR	ON	Olaratumu-
CD3CD3	TRON	~~Basilix-~~, Muromo-, Blinatumo-	GD	ON	Dinutuxi-	PD-1R	ON	Cemipli-
CD4	ID	Ibalizu-	GPIIb/IIIa	CV	Abcixi-	PD-1R & PD-2R	ON	Nivolu-, Pembrolizu-
CD11	IM	Efalizu-	HER2 neu	ON	Trastuzu-, Pertuzu-, Ado-trastuzu-	PD-L1	ON	Atezolizu-, Durvalu-, Avelu-
CD19CD20	ONIM	Blinatumo-Rituxi-	IL-2R a	TR	Basalixi-	PCSK9	CV	Alirocu-, Evolocu-
CD20	ON, ~~Rh~~	Rituxi-, Obinutuzu-, Ocrelizu-, Tositumo-, Ibritumo-, Ofatumu-	IFN-gam.	H	Emapalu-	RankL	ON, Gyn	Denosu-
			IgE	IM	Omalizu-	RSV	ID	Palivizu-
CD22	ON	Inotuzu-, Moxetumo-	IL-1 Beta	IM	Canakinu-	SLAM7	IM	Etolizu-
CD25p55	TR, N	Dacilizu-	IL-4RIL-5RA	IMP	Dupilu-Benralizu-	TNF-A	IM	Adalimu-, Certolizu-, Inflixi-, Golimu-
CR30	ON	Brentuxi-	IL-5	IM	Mepolizu-, Reslizu-	VEGF	ON	Bevacizu-
CR33	ON	Gemtuzu-	IL-6R	IM	Tocilizu-, Sarilu-, Siltuxi-	VEGFA	OP	Ranibizu-, Brolucizu-
CD38	ON	Daratumu-	IL-12	IM	Ustekinu-	VEGFR	ON	Ramucizu-
CD52CD79b	NON	Alemtuzu-Polutuzu-	IL-17A	IM	Ixekizu-, Secukinu-	vWF	H	Ceplacizu-
CGRP	N	Fremanezu-, Galcanezu-	IL-17RAIL-22 alpha	IMTR	Brodalu-Basilixi-	IntegrinA-4 B-1	N	Natalizu-
CGRPRCTLAC. C.diff.Dabigatran	NONIDCV	Erenu-Ipilimu-Bezlotoxu-Idarucicu-	IL-23p	IM	Tildrakizu-, Ustekinu-, Guselku-, Risankizu-	IntegrinA-4 B	IM	Vedolizu-, Natalizu-
						7Sclerotin	Gyn	Romosozu-

Abbreviations: TA – therapeutics areas, CD – Complementary determining region, C.diff. – Clostridium dificilli, CV – cardiovascular, EGFR – epidermal growth factor receptor, GD2 – ganglioside disialo 2, Gen – genetic disease, Gyn – Gynecology, H – hematology, ID – infectious disease, IFN gam. – interferon-gamma, IgE – Immunoglobulin E, IL – interleukin, IM – immune disease (dermatology and/or gastroenterology and/or neurology and/or pulmonary and/or rheumatology), N – neurology, ON – oncology, OP – opthalmology, P – pulmonary, PD-1R or PD-L1 – Proteins on T-cells, PDGF – platelet derived growth factor, PCSK9 – proprotein convertase subtilisen/kexin type 9, Rh – rheumatology, RSV – respiratory syncytial virus, TNF – tumor necrosis factor, TR or Tr – transplant, VEGF – vascular endothelial growth factor, vWF – von Willebrand factor.

Table 3.3 Monoclonal antibody traits (as of December 31, 2019)

Mabs constructs: • Mo – 5 (0% human, 100% murine)) • Xi – 11 (30% human, 70% murine) • Zu – 46 (90% human, 10% murine) • Mu – 8 (100% human) • Lu – 8 (100% human) • Others – 13 **Immunoglobulin (IgG) isotypes:** • IgG1-kappa – 52 • IgG1-lambda – 4 IgG1 unspecified – 4 • IgG2 – 8 • IgG4 – 12 • IgG2/IgG4 – 2 • Unspecified – 8 **Other structural traits:** • Antibody drug conjugates (ADC) - 8 • Biosimilars – 22 • Bispecific mab – 1 • Antibody fragments – 6 • Fusion proteins – 6	**Recombinant host cells – origin:** • CHO – 49 • "Mammalian" – 11 • Mouse myeloma – 15 • NS0 & SP2/0 • *E. coli* – 6 • Unspecified – 10 **Phage display origin: 6** **Uses: [97]** • Oncology – 37 • Gastroenterology – 7 • Dermatology – 6 • Ophthalmology – 6 • Cardiovascular – 6 • Rheumatology – 5 • Infectious Disease – 5 • Neurology – 4 • Hematology – 6 • Other – 15

Abbreviations: CHO – Chinese hamster ovary cells, *E. Coli* – *Escherichia coli* bacteria, IgG – immunoglobulin G isotype, mab – monoclonal antibody, "mammalian" – unspecified mammal species, NS0 – nonsecreting murine myeloma cells, SP2/0 – nonsecreting myeloma cell type.

Historically, the production of mabs most often in biotechnology manufacturing is in a mouse model to create mouse mabs versus human antigens. We are developing a product (mab) to attack and eliminate these target human antigens causing a disease. The mouse will create highly specific mabs against the human antigen (target). Antigens can be freely floating in a physiologic fluid, for example, blood, or most often are attached to the surface of cells. One cell will have many different antigens on its cell surface in both normal cells and disease cells; the disease will usually cause a cell surface to manifest abnormal protein antigens that can be attacked and bound up by mabs. The ideal traits of an antigen enhancing mab attack are manifold, as follows: (1) presence in foreign or abnormal cells and minimally or not on normal cells, (2) high percentage of patients with a disease exhibiting the antigens, (3) homogeneous antigen presence on cell surface of the disease cells, (4) consistent expression of the antigens over time as the disease progresses and is treated, (5) little shedding of the antigen into biologic fluids bathing cells, (6) presence of the target antigen on various disease cells (e.g., various cancers manifest the same antigen such that one mab can attack multiple cancers).

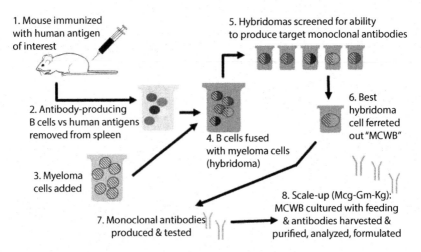

Figure 3.2 Development and manufacturing of monoclonal antibodies (hybridomas).

Abbreviation: MCWB – master cell working bank. (Reprinted with permission from Amgen Inc.)

The primary mab manufacturing process is found in Figure 3.2, wherein we create mabs versus an antigen of interest in a disease and create a therapy (a mab) to attack the antigen and help resolve the disease.

In Figure 3.2 in the diagram area #1 above, a mouse is inoculated with human tissue with an antigen of interest related to a human disease. In the mouse (area #2), APCs will pick up the foreign substances, binding the human antigens to their MHC II molecule. APCs then carry them to the lymph system, wherein further binding occurs between the MHC II-antigen complex in CD34+ T-lymphocytes. In turn, this complex stimulates mouse plasma cell creation from B-cells in the spleen. These specific plasma cells produce the specific mabs versus the specific antigens. In diagram area #3, the plasma cells producing mabs are removed and collected from the lymph system and spleen. However, the quantity of mabs (protein) produced by any one mouse or even a very large number of mice is very small. It is not yet a commercially viable manufacturing process. In biotechnology for mabs (diagram area #3), myeloma cells are employed that have the capacities for very large production of proteins (mabs included), and they are nearly immortal cells. In diagram area #4, the myeloma cells are fused with the mouse plasma cells to create murine hybridoma cells. The hybridomas have the combined characteristics of very high production of proteins (mabs as well), long life span, ability to grow in vitro, and production of highly specific mabs against human disease antigens, all of which make mab manufacturing more commercially feasible. Additionally, hybridomas can be patented, offering further patent protection and delaying marketing of copycat products. Diagram area #5 shows that the various plasma cells need to be screened for which mabs they produce, in order to single out the plasma cell that makes the desired mab. As indicated by diagram area #6, the best hybridoma cells are found related to the cell productivity and quality of mabs. These hybridoma cells are the new "master cell working bank" (MCWB) used to produce large amounts of specifically targeted mabs. Diagram area #7 states next that the mabs are produced and fully characterized and tested. The final diagram area, #8, enumerates that scale-up is conducted for the desired mab production, similar to recombinant protein manufacturing with host cells and mammalian fermentation, purification, formulation, and product analyses.

An alternative mode of development for monoclonal antibodies is antibody "phage" display (APD), which has been used in 5 of the 86 mab molecules currently marketed (adalimumab, belimumab, necitumumab, ramucirumab, raxibacumab). In APD, mabs are fully developed in the test tube, based on the physical link between function (antigen binding of mabs) and information (mab genes) in nanoparticle phages (viruses that infect bacteria, such as *Escherichia coli*). The high antigen affinity and high specificity of mabs to their antigen allows APD to create the specific mab for the specific antigen of a disease. Four main elements are required, that is, (1) preparation of human antibody gene libraries containing millions of mab genes, (2) genetic engineering of bacteriophages to bind mabs [the antigen-binding fragment (Fab) or single-chain variable fragment (scFv) segments of a mab] and serve as probes to filter out mabs against desired antigen targets, (3) serial panning of the gene libraries with the bacteriophage-mab probes, (4) analysis of mab clones. Then, the manufacture of mabs by recombinant DNA methods can ensue. An advantage of this APD system is that the mabs are fully human. A limit to this set of procedures is the laborious nature of the process.

Monoclonal antibodies are more powerful molecules in therapy than most biologics or drugs, directly related to two sets of parameters. First, a mab molecule possesses high specificity for disease antigens and high binding affinity to receptors. Second, their multiple mechanisms of action against the offending molecules or cells are especially advantageous. Four distinct mechanisms are possible modes of action: (1) binding to cells inhibiting their activity, arresting disease progression, for example, IgG-E antigen on mast cells in eosinophilic asthma; (2) binding to a cell receptor, blocking activation on the cell surface receptor and arresting disease manifestation, for example, HER2 oncogene receptor in breast cancer; (3) binding to freely floating ligands (that are antigens), preventing ligands from attaching to their cell receptor, diminishing disease activity, for example, tumor necrosis factor (TNF) in immune diseases; (4) binding of Fc section of mabs with WBCs, which are involved in immune defense, causes activation of an immune reaction cascade of several possible types; ADCC (mab coats disease cells and attracts WBCs to kill abnormal cells), CDCC (complement protein activation with cytokine release to foster cell killing and further activating other cell killing), or PDCC (phagocyte activation and killing offending cells), for example, HER2 oncogene again by the other mechanism of cell cytotoxicity. One mab can have several mechanisms of action occurring simultaneously against an antigen and disease to ameliorate the disease.

The murine origin of mabs creates significant limits for these products because administration of murine mabs (murine proteins) to humans leads to the body's rejection against the foreign murine nature of the protein, an immune rejection reaction. Limitations of murine mabs are twofold; (1) a common adverse reaction, that is, a toxic human anti-mouse antigen (HAMA) response with fever and chills, and (2) less binding of the mouse mabs to the human target cell, thus limiting their activity and product efficacy. A human's immune system can react with various functional types of mabs being generated against the therapeutic mab molecule being administered (considered foreign protein by the immune system), such that the mab will be neutralizing, or clearing, or sustaining, or causing HAMA, or causing allergic reactions, or do nothing. Furthermore, immunogenicity against mab molecules is common and varied between patients. One study presented how common immune reactions to therapeutic mabs are, that is, 89% of 121 mabs displayed some type of immune reaction with 49% impacting efficacy. A patient's genetic make-up, current immune status (normal or compromised), and the immune involvement of the disease being treated will impact immunogenicity and subsequent reactions, as well as the mab type of protein and its formulation. Therefore, scientists now manipulate mabs by substituting human subunits for the four murine subunits, creating chimeric molecules, part murine (about 25%) and part human (about 75%). Even more humanization can be achieved to about 90%, called "humanized" mabs, with only the CDRs being murine (complement determining regions that identify and bind to antigens). Furthermore, mice have been bred by genetic engineering to produce fully human antibodies, by knocking out the primary four murine genes and replacing them with human genes that then will make fully human proteins in mice. This humanization of murine mabs lessens toxicity and results in an increased activity. Further extensive molecular engineering of mabs with a host of constructs and outcomes will be covered in the molecular engineering section of this book.

Another type of adverse reaction with mabs can be an acute allergic reaction, manifesting as a local effect at the injection site up to, rarely, anaphylaxis. Susceptibility to infections can occur with the mabs that have a mechanism to block the immune system, such as adalimumab and infliximab used to lessen the immune system in immune diseases, causing, infrequently, especially unusual and severe infections such as fungi, viruses, and tuberculosis. An individual mab can possess its own unique adverse reaction, also called collateral damage, such as cardiac dysfunction with trastuzumab or tumor lysis syndrome with rituximab. Finally, an additional reaction with mabs is disease exacerbation upon discontinuation of the mab.

chapter four

The science of biotechnology

Peptides, vaccines, cell and tissue therapy, liposomes, nanobiotechnology, animal-sourced products

Peptides

Product development for peptides has been very successful over 40 years. The total number of U.S. Food and Drug Administration (FDA)-approved peptides is 110 products and 93 distinct molecules as of December 31, 2019. A peptide is comprised of a sequence of amino acids from as little as 4 to as many as 100, after which the molecule is considered a protein based on molecular size (number of amino acids) and its complexity (domains, glycosylation, three-dimensional (3D) configuration). Peptides fill a niche as smaller than proteins, but larger than small molecules (drugs), along with possessing the pharmacologic activity of the protein. The development of peptide therapeutic products can involve finding a peptide in nature with the desired properties in humans, then production follows either recombinant DNA (rDNA) manufacturing or *de novo* synthesis (molecular engineering) of a peptide and the screening for its activities. The development and manufacturing methods for peptides are rDNA technology for 46 products and molecular engineering/synthesis for 62 products. The early development of peptides focused on naturally occurring endogenous molecules, often hormones and ligands, such as oxytocin and insulin. Analogs of existing hormones such as octreotide versus somatotropin were created, and then medicinal chemistry came to the forefront in development. Also, animal sources have been utilized to obtain peptide products to be used in humans, for example, exenatide from Gila monsters for diabetes mellitus, bivalirudin from leeches as a thrombolytic, and eptifibatide from snakes' venom as another thrombolytic. The traits of peptides are enumerated in Table 4.1, listing the number of products from animal sources, recombinant versus synthetic products, host cells, types of molecular engineering, and size of products based on number of amino acids.

Limits to peptide usage are their relative short half-lives due to extensive *in vivo* degradation by the ubiquitous peptidase enzymes, relatively

Table 4.1 Traits of peptide products (as of December 31, 2019)

• Animal/environment source discovery	10
• Recombinant products:	47
Host cells:	
• *E. coli*	21
• *S. cerevisiae*	19
• Other	7
• Synthetic products	63
• Molecular engineering:	74
• Polymers/depot	8
• Cyclic molecules	6
• Fusion proteins	5
• Pegylation	3
• Number of amino acids: [4–101]	
• Less than 10	46
• 11–30	17
• 31–70	43
• 71–101	4

poor stability, injectable administration, and inability to cross cell membranes. Product delivery of peptides was improved with special formulations, such as amino acid manipulations, conjugation of peptides, pegylation, polymers, and depot formulations, designed to improve the stability and half-lives of peptide products, as listed in Table 4.1. The targets of peptides in their mechanisms of action are a variety of extracellular target receptors, such as G-protein coupled receptors, cytokines receptors, antimicrobials, adhesion molecules, and ion channels. The clinical uses for peptides are quite manifold, 84 in number in 16 medical disciplines, especially endocrinology, gynecology, oncology, and cardiovascular areas. Since 2010, peptide development has grown and resulted in 42 additional commercially available products.

Vaccines

Biotechnology involvement for vaccines includes two technologies – rDNA technology to manufacture the infectious antigens in question and/or molecular engineering with fusion molecules of a couple of active components – both to stimulate the human body to create the immunity desired for vaccines versus infectious agents. Biotech vaccines number 31 as of the end of 2019. Traditional vaccines employing egg culturing without rDNA or molecular engineering are not covered here. Vaccines for infections are very different from the therapeutics with drugs and biologicals, in that, they are preventative, eliciting an immune response to ward off future infections. Vaccines must be administered and allowed to establish an effect prior to exposure to an infectious agent. Pasteur's principle for vaccines from the 1870s still holds true today: find *Cause* (usually bacteria or virus), *Isolate* responsible organism and best antigen(s), *Inactivate* organism (from causing an infection [antigen in organism]), *Inject* the vaccine (antigen) to establish the immunity to prevent the infection, *Patient's immune system* creates the immunity versus the infection (creating mabs and/or specific T-cell lymphocytes against the infectious agent). Hence, product development for vaccines involves first finding and isolating the responsible organism for an infection and, especially, the best antigen from the organism to stimulate the strongest and longest acting immune response. Currently approved biotech vaccines prevent infections with the following 11 infections: bacteria – *Bacillus anthracis* (anthrax), *Borrelia burgdorferi* (Lyme disease) (product withdrawn), *Neisseria meningiditis* (meningitis), *Neisseria meningiditis* Group B (meningitis Group B), *Streptococcus pneumoniae* (many serotypes, over 14); viruses – flavivirus (Dengue, 4 serotypes), hepatitis B, human papilloma virus (HPV – 9 serotypes) (cancer prevention), herpes simplex virus with the granulocyte-macrophage colony stimulating factor (GM-CSF) gene (cancer therapy), influenza (various serotypes, 3–4 annually); and protozoa – *Plasmodium falciparum* (malaria).

For immunity, the monoclonal antibody and/or T-cell response in a patient must be sufficient to prevent the organism or virus from establishing itself and causing an infection. Therefore, success in product development is assessed by the immune response ("Immunity") being tested with, for example, sufficient antibody titers in the blood. Furthermore, vaccine development involves identification of the best antigen from the infectious agent, how much antigen to include in the vaccine, inclusion of multiple antigens if needed (more than one species or serotype, more antigens = multiple serotypes), consideration of antigen drift (change in antigens over time), and geographic variations in organisms and antigens. Antigens from offending organisms/viruses can be live-attenuated, inactivated, synthetic, virus-like particles, polysaccharides subunits, or recombinant entities. To establish optimal immunity in clinical trials and for patient care, further work is required, that is: (1) How much vaccine is given? (2) Does it need an adjuvant added as a booster included with the antigen to enhance immune responses? (3) Are second and third vaccine injections necessary, and at what frequency, to optimize immunity? (4) Does the use of the vaccine in a population of people provide added "herd immunity" in close associates? (5) What is the extent and type of immune responses to the vaccine? (6) What is its persistence of protection provided by the vaccine? (7) Given that vaccines are employed commonly to prevent childhood infections, often in newborns, children are involved as subjects necessitating extra caution and observation in clinical trials to establish efficacy and especially safety. Vaccines are considered as very safe therapeutic products, with mostly local irritation and mild allergic reactions. Serious problems are very rare, for example, a compilation of 1000 studies of vaccines involving many thousands of patients found only 14 significant reactions, anaphylaxis, febrile seizures, and encephalopathy. The manufacture of vaccines involves cell cultures of the infectious agent in appropriate growth media, extraction of the organism and, ultimately, the target antigen in addition to formulation of the agent and antigen in stable and nontoxic media, establishing its potency and showing its purity, lack of contamination, and inability to cause infections, which would require extensive testing. Table 4.2 lists 17 such considerations in the development of a vaccine.

Tissue therapy

In tissue engineering in biotechnology, the goal is to create biological substitutes of a tissue to restore, maintain, or improve tissue and organ function. We generally need tissue or cells from patients or normal donors through biopsy, a process for *ex vivo* cell expansion, often a scaffold on which to grow the cells, and a bioreactor system in which the new tissue can grow to its full size, normal structure, functional capacity, and devoid of toxicity, immunogenicity, or fibrosis. In late 1990s, the first tissue/cell therapy products were developed for skin damage or injury and also cartilage damage. Foreskins of newborns were collected and placed in a 3D construct that permits growth of two layers, dermis (fibroblasts) and

Table 4.2 Considerations in development and use of vaccines

- Organism (antigens) identified (1 or more)
- Organism (antigens) isolated (1 or more)
- *In vitro* cell culture of organism for manufacture
- Antigen type, e.g., protein, polysaccharide, inactivated, particles, recombinant, synthetic
- Antigen/organism inactivation (to prevent infections by it)
- Adjuvant inclusion to enhance immunity
- Formulations ingredients (diluents, buffers, stabilizers)
- Testing for immunity (extensive & sophisticated)
- Adverse events – local, systemic, neurologic
- Immune response: mabs vs. T-cells (or both)
- Extent & persistence of immunity
- Dosage of vaccine (amount of antigen)
- When dose given
- How often given
- Interval between doses
- Herd immunity (to be developed in population over time)
- Antigen drift & vaccine antigen changes

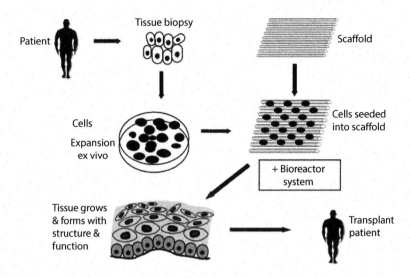

Figure 4.1 Tissue engineering in product development. (Reprinted with permission from Amgen Inc.)

epidermis (keratinocytes), along with a usually nontoxic bioabsorbable scaffold. As the skin substitute grows to become functional skin, secretion of dermal collagen, matrix proteins, cytokines, and growth factors occurs. This construct is used later in venous leg ulcers or burns to accelerate wound healing, thereby reducing health care needs. This process of tissue development for therapeutic use is shown in Figure 4.1.

Also, chondrocytes responsible for cartilage production can be taken from a patient's knee that has serious damage and is repairing poorly. These chondrocytes are then manipulated *ex vivo* with expansion and returned to the patient to normalize cartilage production. In addition, mesenchymal stem cells can be harvested from human chorionic placental tissue, combined with stromal cells and a collagen membrane for wound healing.

Liposomes

A liposome is a spherical phospholipid product comprised of one or usually two concentric circles of lipid molecules with a nonlipid aqueous center. The lipids utilized can vary including phosphatidyl choline (most common), phosphatidyl ethanolamine, or cholesterol. The properties are manifold and advantageous to improve drug delivery and safety, including biocompatibility, biodegradability, hydrophobicity and hydrophilicity, positive- or negative-charged lipid molecules, plus other designed traits, such as, small size down to nanomolecules (about 0.01–0.1 µm = 10–100 nm), carrier potential for drugs and biologics, avoidance of drug degradation within the liposome, sustained release of contents, delivery of contents to cells, and intracellular delivery of contents. Four possible therapeutic goals for liposomes are enhanced efficacy, less toxicity, improved drug delivery, and sustained release of contents.

Figure 4.2 displays the structural elements of a conventional liposome: a bilayer of phospholipids, positive- or negative-charged lipid molecules, with a hydrophilic (aqueous center) and hydrophobic (lipid bilayer) loading sites for drugs. Pegylation of a liposome creates a stealth mode wherein the liposome is less well recognized by the immune system as a result of the addition of polyethylene glycol compounds to the outer layer of the liposome. Liposomes can be designed to target a desired site through varied attached species, such as, monoclonal antibodies, peptides, proteins, carbohydrates, or small molecules. At its site of action, a liposome can be taken up by a cell via endocytosis with release of the contents via fusion of a liposome at the target cell membrane or through phagocytic activity of macrophages. Liposomes currently include the following types of products: antifungals, antimicrobials, cancer chemotherapy, analgesics, and anti-inflammatory steroids.

Nanobiotechnology

The field of nanobiotechnology involves the use and study of particles, organelles, instruments, drugs, and devices that measure or function within a size parameter of 1–100 nm. A nanometer is one billionth of a meter in size (10^{-9}). To give these sizes some relevance, the red blood cell size is about 5 µm (5000 nm), and an aspirin molecule is less than 1 nm. Biologic applications are still at an early stage of scientific evolution including the areas of bioanalysis, drug delivery, therapeutics, biosensors, medical devices, and tissue engineering. Drug discovery and diagnostics are the two most significant areas of research in the private sector for nanobiotechnology. Currently, the scanning probe microscope is a tool in common use for cellular study, functioning on a nanometer scale. Nanoparticles can provide new labeling technology in cell or molecule analysis based on the change in color with change in particle size. Use as contrast agents in X-ray imaging is an application, possibly with better image resolution, tissue targeting, and retention in blood. Nanoparticles can be carriers for drugs in very minute amounts, possibly enhancing drug penetration across membranes, changing drug solubilities, and altering pharmacokinetics. Liposomes can function as nanoparticles. The commercially available product Abraxane binds the cancer drug paclitaxel to albumin, which can then be considered a nanoparticle biological drug formulation.

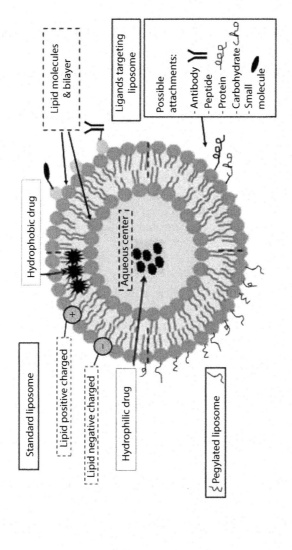

Figure 4.2 Basic structure of liposomes. (Reprinted with permission from Frontiers Media S.A. © 2007–2015.)

Animal and marine products

Identification of pharmacologic activity of biological products in animals for human applications has been a process for a very long time in drug discovery; of course, porcine and beef insulin were extracted from the pancreas of animals and used in patients for most of the twentieth century. Then, in 1982, a recombinant form of human insulin was created and marketed superseding the need for animal insulin. Animal proteins may differ in their construct from human proteins, yet positively present useful pharmacologic properties that can then be used in humans as therapeutic agents. However, they are foreign proteins or peptides to humans with possible immunogenic reactions, such that they can be identified, characterized, and then reproduced employing recombinant technology to humanize them. Thus, we now have various biological products isolated and identified from animals that are used to treat human diseases. From leeches, we have two thrombolytic enzymes, bivalirudin and lepirudin. The Gila monster provides a peptide, exenatide, for glucose control in treatment of diabetes mellitus. The marine snail is a source for identifying a peptide, ziconatide, for severe pain. Snakes from their venom have given us a peptide enzyme, eptifibatide, to be used for thrombolysis in cardiovascular conditions such as acute coronary syndromes and angioplasties.

High-throughput screening

High-throughput screening (HTS) is a technology intended to obtain faster, more high-quality product leads from large libraries of genetic or peptide molecules for physiologic targets for a disease. Lead optimization in product discovery can be incorporated into HTS. Screening can also be done for disease targets. HTS has improved the number of molecules that can be screened for activity by 10- to 1000-fold in much less time. The process of HTS is dependent on improved analytical processes (better surface chemistry, capture agents, and detection methods), miniaturization of equipment, and automation, as well as optimized libraries to be screened. Currently, over 100,000 samples can be tested in a day.

Combinatorial chemistry

Combinatorial chemistry involves the use of the basic building blocks in biochemistry, either the 20 amino acids or 4 nucleic acids, to build new molecules. All the different combinations of a number of building blocks can be created, for example, the use of 10 different amino acids can result in over 3.5 million decapeptide compounds. Huge libraries of compounds are produced, which require screening through HTS and the use of informatics to help sort out the structures and, especially, the activities of all such new compounds.

Proteomics

The proteome is the complete protein makeup in the human body. Proteomics is the study of protein structures and their properties. The proteome is more complex than the genome when we consider the greater complexity of proteins, for example, 20 amino acids versus 4 nucleic acids, and their manifold structural requirements, including the amino acid sequence, disulfide bridges, glycosylation of proteins, the complex carbohydrate structures, the amino and carboxyl ends of proteins and their variation, the isoforms of the same protein in one patient and between patients, and the 3D configuration (folding) of proteins. Proteins have a certain mass, isoelectric point, and hydrophobicity, impacting their activity. Protein function will also potentially change in several circumstances, for example, during development from the fetus to a child to an adult, in disease versus normal physiology, during inflammation versus none, and possibly have different actions at specific tissue sites. All of these different properties will require sophisticated and sensitive analytical technologies to identify and understand protein structure and function. Proteomics assists us in finding new disease targets and possible biological products for therapy.

Cell receptors, signal transduction and protein kinases

Over the last few decades, cell function has been and is being more fully elucidated, and a large group of protein enzymes, protein kinases (PKs), has been found to play principal roles in the communication between and within all cells, resulting in activation of all cell functions. Several thousand PKs exist and are very specific to certain cells and cell functions. Aberrant or excessive cell activity can be mediated by the PKs, contributing to diseases such as cancers or inflammatory conditions. PKs become targets for drug intervention to turn off or reduce their activity and moderate a disease. They are quite desirable targets for a host of reasons and properties given their (1) universality in cells, (2) very high specificity to a cell function, (3) involvement in many principle cell functions, (4) their specific protein structures, (5) the identifiable mechanisms of action within their substructures, and (6) their accessible sites for drugs and biologicals. For example, tyrosine PKs serve as cell receptors and allow binding to a cell of various cell-to-cell communication ligands, such as hormones or growth factors that lead to intracellular auto-phosphorylation of the PK receptor and activation of intracellular PKs, resulting in targeted cell changes. The drug imatinib (Gleevec) was the first PK inhibitor approved for use and treats chronic myelogenous leukemia with a bcr-abl single nucleotide change that creates the Philadelphia chromosome-positive

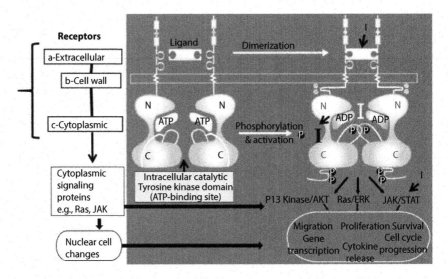

Figure 4.3 Cell receptors – construct and actions.

Abbreviations: ADP – adenosine diphosphate, AKT – protein kinase B, ATP – adenosine triphosphate, ERK – extracellular signal-regulated kinase, I – inhibitor, JAK –
Janus kinase, P – phosphate, Ras – ras proteins (facilitate cell division),
STAT – signal transducer and activator of transcription. (Reprinted with permission from Amgen Inc.)

abnormality. Imatinib disrupts the function of the intracellular domain to
turn off the abnormally functioning cell.

Figure 4.3 presents a diagram of a cell receptor (a very complex multifaceted structure) that is found on the cell wall, which is responsible for
the activation of the cell to perform a particular function. Cell receptors
are stimulated by the binding of a protein or drug ligand to the receptors.

Receptors are complex proteins on the cell surface comprised of three
segments or domains: (1) an extracellular domain, with a variety of binding sites for several various ligands to be attached (represented by different shapes on this domain, loops, diamonds, and rectangles); (2) a cell wall
domain; and (3) an intracellular cytoplasmic domain, a complex structure
that upon activation self-phosphorylates and in turn activates a set of
intracellular signaling proteins. Upon a ligand attaching to a receptor, the
receptor dimerizes, activates, and self-phosphorylates key segments of the
receptor via adenosine triphosphate (ATP) turning into adenosine diphosphate (ADP) and providing a phosphate molecule. Signaling proteins are
activated in the cytoplasm, and in follow-up travel to the nucleus, turn
it on, and initiate a cell action; that is, they communicate with the cell
nucleus to initiate an action for the cell to perform, such as migration,
proliferation, survival or apoptosis, gene activation, cytokine release, and

cell cycle progression. Inhibitors of cell function (as shown in Figure 4.3 as a capital "I") can disrupt the function of a receptor in at least three points: through (1) ligand binding to the receptor, (2) phosphorylation interruption, and (3) blocking the action of cytoplasmic signaling proteins.

Cell therapy

In biotechnology, cell therapy involves obtaining healthy cells from a specific tissue, selecting out a specific subset of cells with certain desirable properties, and enhancing the activity of these cells through *ex vivo* manipulation. We then return these specifically selected, enhanced, and activated cells to patients who have damaged tissue and whose cells are not sufficiently functional, thereby ameliorating a disease. The patient also may benefit from enhanced cells improving disease response. For example, bone marrow progenitor cells can be collected from peripheral blood, bone marrow cells, or cord blood of cancer patients, and the cells with the greatest regenerative potential are selected and separated through various cell-tagging processes followed by *ex vivo* cell expansion. Selection also eliminates blood-borne cancer cells. Following life-threatening high-dose chemotherapy to kill cancer cells in a cancer patient, which destroys almost all the bone marrow hematopoietic cells, these selected progenitor cells are administered to the patient to accelerate regeneration of bone marrow and white blood cell production, especially preventing infections. See Figure 4.4 for the full process of cell therapy in oncology.

Figure 4.4 presents the dual processes of peripheral blood progenitor cell preparation (*in vitro*, Steps 1–4) and usage in a cancer patient (*in vivo*, Steps A–D). First (Step A), a patient receives a CSF and/or low dose chemotherapy to stimulate the bone marrow to increase progenitor cell production into the bloodstream. Second (Steps B and 1), the blood is harvested from the patient including normal blood cells, tumor cells, and the desired progenitor cells. Third (Step 2), the progenitor cells, recognized by the appearance of the CD34 antigen on their cell surface, are labeled with a magnetic label. Fourth (Step 3), a magnet is employed to separate and attract the CD34 progenitor cells, removing the tumor cells and most normal blood cells. Fifth (Step 4) the CD34 pool of progenitor cells is expanded greatly in number using a mixture of growth factors (possibly, G-CSF – granulocyte-CSF, GM-CSF – granulocyte-macrophage-CSF, SCF – stem cell factor, MGDF – megakaryocyte growth and development factor, IL-3 – interleukin-3, and IL-6 – interleukin-6). Sixth (Step C), while preparing the progenitor cell therapy in the laboratory, the patient receives high-dose chemotherapy that obliterates the bone marrow of cells. Seventh (step D), the patient receives the live-saving dose of peripheral blood progenitor cells.

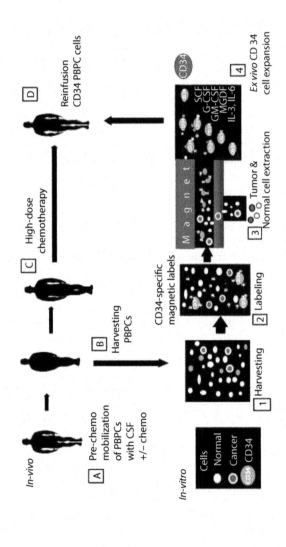

Figure 4.4 Cell therapy in bone marrow transplants and chemotherapy.

Abbreviations: CSF – colony stimulating factor, G – granulocyte, GM – granulocyte-macrophage, IL – interleukin, MGDF – megakaryocyte growth and development factor, PBPC – peripheral blood progenitor cell, SCF – stem cell factor. (Reprinted with permission from Amgen Inc.)

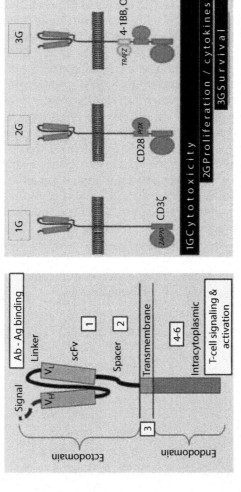

Figure 4.5 T-cell enhancement in immuno-oncology.

Abbreviations: Ab – antibody, Ag – antigen, scFv – single-chain variable fragment, V_H – variable heavy chain, V_L – variable light chain, 1G – first generation, 2G – second generation, 3G – third generation. (Reprinted with permission from Amgen Inc.)

Cell therapy and immuno-oncology

Cell therapy has achieved a new level of sophistication and efficacy as an immuno-oncology treatment in conjunction with chemotherapy to treat cancer. In chimeric antigen receptor T-cells (CAR T-cells), the immuno-logic functionality of immune T-cells is substantially improved to attack cancer cells. A patient's T-cells are harvested. Then the T-cells are manip-ulated *ex vivo* with receptors modified to specifically and better target cancer cell antigens, as well as also enhancing the cell killing function of the T-cells. This manipulation with added functions being incorporated into the T-cell receptors is effected through gene manipulation to direct the cells' protein manufacturing of receptors to add in these additional functions. New genetic material is carried into the T-cell via a transfec-tion process. These CAR T-cells with highly specific and elevated func-tionality are returned to the patients to significantly improve the patient's immune function to enhance cancer therapy. Figure 4.5 shows in the left frame the cell receptor's three domains, with the extracellular segment (also called ectodomain) an antibody fragment for cell binding to a tar-get, a spacer section manipulated to attach to the cell wall segment of the receptor (transmembrane section), and several possible manipulations to the intracytoplasmic segment (endodomain) of the receptor, impact-ing T-cell signaling and cell activation. The right frame of the figure is a cartoon example suggesting three structural manipulations of the cyto-plasmic segment: first-generation genetic/protein change for cytotoxicity improvement, second-generation change for additional proliferation and cytokine improvements, and third-generation change adding feature of increasing the survival of the T-cell lymphocytes.

chapter five

The science of biotechnology
Molecular engineering (proteins, monoclonal antibodies, biosimilars)

Over half of all biotech molecules are molecularly engineered in some fashion. Molecular engineering now is performed commonly *de novo* up front in the creation of new biotech molecules. Also, the newer generation of biological products are often the derivatives of existing molecules that are altered structurally to enhance or change their properties, resulting in new molecules. Molecular manipulation of biologicals is now a predominant technique used in product development, through 15 such different processes, as listed in Table 5.1.

The properties of proteins that can be altered by molecular engineering are quite extensive, including receptor affinity, receptor selectivity, pharmacokinetics (half-life, metabolism), immunogenicity or antigenicity, effector function, potency, safety (adverse events), stability (*in vivo* degradative enzyme resistance or *in vitro* shelf life), solubility, absorption by new routes of administration, and manufacturing yield. One structural change in a protein likely will result in more than one pharmacodynamic change, possibly with one positive and one negative, with a resulting composite effect. Besides enhanced properties, these new biological molecules have the added benefit of being patentable as new drugs.

Molecular engineering of proteins and peptides

Insulin is a large peptide comprised of 51 amino acids arrayed in 2 chains; A-chain of 21 amino acids and B-chain consisting of 30 amino acids with 3 disulfide bridges (S-S) between the 2 amino acid chains, and a carboxyl (COOH) terminal end and an amino (NH_2) terminal end for each chain, as shown in Figure 5.1. A variety of changes in the peptide structure have been performed to create new insulin-like molecules with new pharmacodynamic and pharmacokinetic properties.

Table 5.1 Types of molecular engineering

Proteins and peptides:
- Amino acid changes (replacement, elimination, or addition)
- Truncation of peptide domains
- Glycosylation additions or deletions
- Pegylation
- Fusion of molecules (combinations of proteins, peptides, antibody fragments, drugs, toxins, and vaccines)
- Isoform selection
- Monomer versus dimer molecule selection

Monoclonal antibody engineering:
- Humanization of antibodies
- Antigen targeting in the complementarity determining regions
- Isotype selection
- Isotype subset selection
- Conjugates of monoclonal antibodies (mabs) and drugs
- Pegylation
- Bispecific antibodies
- Antibody fragments
- Nanobodies

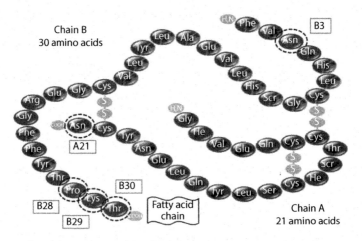

Figure 5.1 Insulin structure. (Adapted with permission from canstockphoto. com.)

Sites of molecular alterations of insulin are identified in Figure 5.1 by a dotted circle around the specific amino acids (five such sites of change), which will be elucidated further below. The peptide sequence of insulin has been altered to create seven different new molecules with varied onsets and/or durations of activity in their glucose mechanism of action (e.g., aspart form, degludec, detemir, glargine, glulisine, glycine, lispro molecules). Insulin aspart has the B28 amino acid changed from proline to asparagine and has a rapid onset of action. The degludec form and the detemir form of insulin have the B30 threonine amino acid replaced with a carbon fatty acid chain (C-16 and C-14, respectively), imparting a long duration of action. The insulin glulisine molecule replaces two amino acids, B3 asparagine to lysine and B29 lysine to glutamine, causing a more rapid onset and shorter duration for its glucose actions. The insulin glargine form has one amino acid change at A21 from asparagine to glycine and two added arginine species to the carboxyl end of the B-chain, imparting a long action to the molecule. Insulin lispro molecule also possesses two amino acid changes, B28 proline to lysine and B29 lysine to proline, resulting in sustained activity.

Protein manipulation for enzyme proteins can serve as a good example of several possible different manipulations and new products. Alteplase is an enzyme protein used for thrombolysis and to prevent death in acute myocardial infarction related to preventing clot formation. The protein is fairly complex with 527 amino acids, 14 disulfide bridges, glycosylation at 3 amino acid sites, and 5 peptide domains, as demonstrated with rectangles in Figure 5.2.

In Figure 5.3, truncation of the amino acid sequence at amino acid 356 (notation 1a, by a lightning bolt in Figure 5.2) eliminates three domains (finger, epidermal growth factor, kringle 1) (notations 1b, with large open crosses), leaving intact the two domains, protease and kringle.

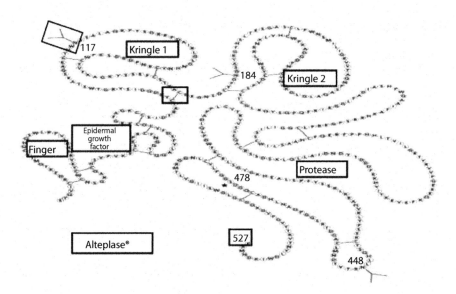

Figure 5.2 Protein structure of alteplase molecule.

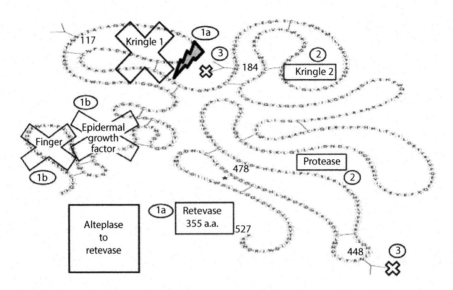

Figure 5.3 Protein molecular engineering – alteplase to retevase.
Abbreviation: a.a. – amino acids. (Reprinted with permission from Amgen Inc.)

These molecular alterations result in a new biological alteration with 355 amino acids and 8 disulfide bridges (reteplase). In addition to eliminating the 3 domains, the 2 remaining carbohydrate (CHO) structures at amino acids 184 and 448 were removed (notation 3, with small crosses), leaving 2 domains [protease and kringle 2 domains (notation 2)], yet maintaining quite similar thrombolytic activity, no immunogenicity, and sustained stability for the molecule.

As displayed in Figure 5.4, another set of changes was made to the alteplase protein, that is, 6 amino acid exchanges at sites #117 and 103 in the kringle domain, and 296–299 in the protease domain, along with added glycosylation at site 103 and deletion of a CHO molecule at 117, resulted in a new protein, tenecteplase (TNKase). All the other main structural features are retained: 527 amino acids, 5 peptide domains, and 14 disulfide bridges. This new protein possesses several advantageous properties: improved administration, that is, a desired intravenous bolus instead of infusion during an acute medical episode, more fibrin specificity, less degradative enzyme susceptibility, longer half-life, and no added toxicity or immunogenicity, creating a superior pharmacologic and pharmacokinetic profile. Such molecular engineering with multiple and varied structural changes creates a series of experiments that are performed separately and in varied combinations. A large series of new molecules (possibly hundreds of molecules) is created, also necessitating each molecule to be tested for its collective new pharmacologic and pharmacokinetics

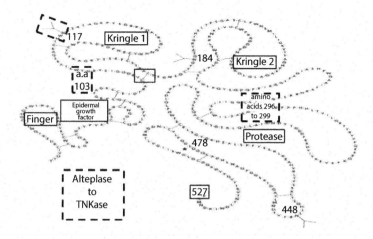

Figure 5.4 Protein molecular engineering – alteplase to TNKase.

Abbreviations: a.a. – amino acids, TNKase – tenecteplase. (Reprinted with permission from Amgen Inc.)

properties. Many serial examinations are needed to ferret out the best new molecule with the best summary of properties that were targeted to be improved.

Coagulation proteins (also known as blood factors) are a series of naturally occurring proteins that are activated in a serial fashion to achieve normal blood clotting and avoid bleeding episodes (Blood Factors 7, 8, 9, 10, 13). These protein molecules are very large, comprised of thousands of amino acids, divided into multiple chains. Therapeutically, they are administered to treat a specific deficiency of a blood factor, which occurs in hemophilia (e.g., hemophilia A – Factor 8 deficiency, hemophilia B – Factor 9 deficiency). For coagulation proteins, Factor 8 (a pro-protein) contains 2,332 amino acids in a single continuous chain comprised of 6 domains plus 3 small linker chains, [A1 – (a1 linker) – A2 – (a2 linker) – B – (a3 linker) – A3 – C1 – C2]. The 3 A-chains are heavy chains with about 329–371 amino acids each; the B domain is a heavy chain with about 900 amino acids; and the C-chains are light chains comprised of 150 and 160 amino acids, respectively. The 3 linker chains contain about 42 amino acids each. During normal physiologic activation of Factor 8, the B domain is deleted. In molecular engineering, 10 of these coagulation proteins have some form of truncation of almost all or parts of the B domain, yet maintain full functional capacity, although different durations of action and molecular stability (Original Factor 8 molecules – Recombinate in 1992 and Kogenate in 1993; Blood Factor 8 derivatives – Advate, Afstyla, Esperoct, Jivi, Kovaltry, NovoEight, Nuwig, Obizur, ReFacto, Xyntha over the 17 years of 2003 to 2019).

Glycosylation

In the human body, proteins are all glycosylated, possessing carbohydate (CHO) molecules in their structure, which are comprised of a combination of several different CHO species, that is, mannose, or fucose, or galactose, or sialic acid, or *n*-acetylglycine. The CHOs are combined and configured as shown in the cartoon diagram (Figure 5.5) in a candelabra-like formation. However, many proteins can remain functional without the CHO content. Some proteins must have the glycosylation content the same as created naturally in mammals to be a fully physiologically active molecule, for example, epoetin alfa for red blood cell production in bone marrow. The impact of the CHO on a protein can be extensive, with changes in protein folding; ligand/receptor binding; biological activity, stability, or solubility of the molecule; protection from enzyme proteolysis; half-life of the molecule; and immunogenicity. The glycosylation entails changing or adding CHO species, for example, sialic acid residues, to the amino acid backbone of a protein. A CHO molecule is attached to an oxygen or nitrogen species in the protein structure and only at the amino acid sites in the protein where a series of three adjacent amino acids (asparagine-serine-threonine) exist. The CHO molecular engineering also requires site-directed mutagenesis (genetic engineering to alter the amino acid sequences), along with a specific transferase enzyme in the endoplasmic reticulum and chaperone proteins engaged as well. Figure 5.5 displays increased glycosylation of the epoetin alfa molecule to create darbepoetin alfa, both utilized to correct the anemia in chronic renal disease and some cancers. The molecular changes and the net outcomes are outlined in Figure 5.5.

Figure 5.5 presents four individual CHO species in one structure in a specific stick-form configuration, like a candelabra form. Epoetin alfa possesses four CHO structures attached to the protein backbone, while the molecular engineered molecule darbepoetin alfa has two additional CHO

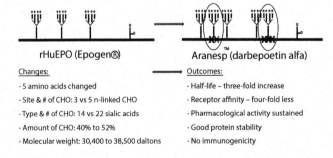

Figure 5.5 Glycosylation of proteins – epoetin alfa to darbepoetin.
Abbreviation: CHO – carbohydrate. (Reprinted with permission from Amgen Inc.)

structures (the "x" in the diagram indicates amino acid changes and the site of addition of CHO). The altered CHO structure (hyper-glycosylation) in the protein, epoetin alfa, created the biological, darbepoetin alfa, which has an extended duration of action that allows less frequent administration along with the same pharmacologic action of stimulating bone marrow stem cells to produce the colony stimulating factor, erythropoietin.

Pegylation

Pegylation is a process in which polyethylene glycol (PEG) molecules are added to the protein structure. It has been done for eight different types of molecules: interferons (e.g., interferon-2a), a growth factor (e.g., filgrastim), all liposomes (e.g., doxorubicin), three enzymes (e.g., pegasparaginase, pegademase, and pegloticase), three coagulation proteins (e.g., Jivi), a hormone (e.g., pegvisomant), an aptamer (e.g., pegaptanib), and an antibody (e.g., certolizumab pegol). Figure 5.6 presents a diagram of a pegylated molecule, peg-filgrastim compared to its original molecule, filgrastim.

Pegylation will vary in several ways: (1) in the number and types of PEG molecules, (2) in the attachment site in the protein chain (only attached at three specific amino acids, cysteine, histidine, or lysine), (3) in the number of attachment sites on a protein, and (4) in the type of linker molecule for the PEG linkage to the protein. Any of these variations will alter the pharmacokinetics or extent of pharmacologic activity of the protein. A reduced toxicity and antigenicity with sustained stability and solubility are desired outcomes with pegylation. The addition of PEG may extend the product's half-life or duration of effect: alpha interferon-2a (Pegasys) was pegylated at nine sites versus filgrastim (Neulasta) at one site, and both slowed the clearance of the molecule from the body and extended the half-life, yielding sustained blood levels, permitting much less frequent dosing, yet the desired activity of the molecule was

PEG   +   Protein   + Linker segment

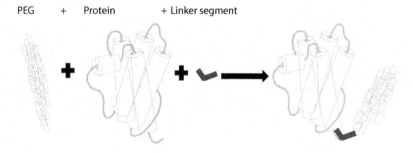

Figure 5.6 Pegylation of proteins – filgrastim to pegfilgrastim.
Abbreviation: PEG – polyethylene glycol. (Reprinted with permission from Amgen Inc.)

maintained. Additionally, and separately, pegylation of large biologic particles can afford a measure of protection against immunogenicity toward a molecule, as in the case of most liposomal products.

Monoclonal antibody engineering

Monoclonal antibodies (mabs) are structurally highly complex molecules with 16 domains in a "Y" shape configuration, all of which can be modified; 8 constant and 4 variable domains, 4 light and 8 heavy chains, 4 disulfide bridges, 2 CHO structures, 4 complement determining regions for antigen binding, and the constant heavy Fc section, possessing several cell cytotoxicity functions. Mabs always necessitate some form of molecular engineering to create a therapeutically useful and optimal molecule, involving 11 possible engineering alterations: (1) antigen targeting of the complementarity determining regions (CDRs), (2) isotype choice (IgA, -D, -E, -G, -M), (3) IgG subtype choice (IgG1, -2, -3, -4), (4) humanization (partial up to 100%), (5) conjugation, (6) bispecific mabs, (7) glycosylation, (8) fusion proteins, (9) antibody fragments, (10) pegylation, and (11) nanobodies.

Humanization of monoclonal antibodies and conjugates

Figure 5.7 presents the molecular engineering for the humanization of murine mab molecules in three possible ways. The first change is listed as #1 in the figure, "chimeras," and involves only changes in the CDRs

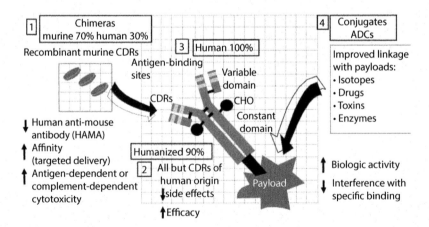

Figure 5.7 Monoclonal antibody engineering – humanization and conjugates.
Abbreviations: ADC – antibody drug conjugate, CDR – complementarity determining region, CHO – carbohydrate. (Reprinted with permission from Amgen Inc.)

responsible for target/antigen binding [30% human, 70% murine molecule]. The second change is listed as #2 in the figure, "Humanized" forms [90% humanized]. The third change is listed as #3 in the figure, "human" mabs [100% human]. The goals of humanization are reducing or eliminating the human anti-mouse antibody rejection of murine mabs, less adverse effects from mouse antigens, improved affinity of the mabs (CDRs) to target cell surface antigens, and improved mab mechanisms of action for cytotoxicity (both antigen-dependent and complement-dependent types, ADCC and CDCC, respectively, at the C_H3 Fc segment of the mab). Over 95% of mabs are humanized to some extent.

Figure 5.7 also indicates the engineering to produce a "conjugate" form of a mab by adding a payload to the mab (section #4 in the figure). Engineering of mabs to serve as carriers of toxic materials involves consideration of the target antigen for the target disease, the type of antibody used as the delivery vehicle, the type of toxic compound ("conjugate") attached to the mab or mab fragment, and the linker technology binding the conjugate to the mab. The target antigen (Ag) often is a cell surface protein that serves as a cell receptor for cell activation, also called "CDR"; any one cell can have hundreds of antigens on its surface. The CDR structure of mabs attaches to these receptors on cell surfaces (antigen sites) in cancer cells and immune disease cells. The ideal traits for an antigen with mab therapy are manifold and many of these following attributes are needed: (1) high specificity by the desired antibody (Ab), (2) with high-binding affinity between them, (3) expression in the tumor or abnormal tissue at high levels, (4) homogeneity at the target cell surface, (5) existence in significant percentages of patients with the disease, (6) expression throughout the disease process, and (7) no or minimal expression in normal tissue. Also, the antibody should possess accessible sites for loading the toxic chemical or product (conjugate) and be able to release the "conjugate" at the target tissue site.

Antibody drug conjugates (ADCs), serving as carriers of cell-killing toxins directly to the target cell, possess several possible cell killing actions; the cell-killing action of the conjugate payload to the mab, along with the mab's inherent cell killing by cell cytotoxicity and also mab binding to cells, fostering immune reactions against the cell–mab complex. The payloads can be toxic chemicals (e.g., calecheamycin, emtansine), cell toxins (e.g., ricin), bacterial toxins (e.g., diphtheria), or radioactive nuclides (e.g., Indium-111, Yrttium-90), producing localized radiation. Conjugation also involves the use of a linker molecule that binds the payload to the mab. The properties and traits of the ADC and its components are listed further in Table 5.2.

Monoclonal antibodies can be engineered further so that a single mab can bind to two different target antigens, bispecific antibodies. Please note again that a mab has two parallel structures side-by-side in the "Y" shaped

Table 5.2 Monoclonal antibody (mab) drug conjugates (ADCs) – properties

Mab and ADC functions, desired:
Normal mab functions to be sustained with a conjugate:
- Mab distribution to desired target cells or organs
- Antigen binding of mab
- High affinity binding to target receptor (antigen)
- Cell internalization of ADC
- Catabolism of ADC complex with the toxin release into target cells
- Mab mechanisms of action vs. target cells sustained, e.g., antigen-dependent cellular cytotoxicity (ADCC)

Payload (toxin) actions:
- Conjugate stability in plasma
- Toxin release in target cell
- Payload action vs. target cell
- Cell (target) disruption/death
- Isotope payloads killing of diseased bystander cells

Problems to avoid or minimize for ADC:
- No adverse pharmacokinetic changes
- No immunogenicity change
- No molecular aggregation (inactivation)
- Normal cell killing
- Isotope payloads killing of normal cells

Desired linker properties:
- Binding of mab to payload (right location, right affinity)
- ADC stable in plasma
- Cleavage intracellularly
- No toxicity to patient
- No immunogenicity

configuration with two CDRs in the variable regions on each side of the Y arms. The CDRs on each half of the mab molecule can be redesigned to bind to different targets in the same mab. See Figure 5.8. The mab can bind both to receptors on a cancer cell and also to a leukocyte involved in the immune system (antigen presenting cells, macrophages, natural killer cells, or T-cells), bringing the cancer cell directly adjacent to immune cells to enhance cancer cell killing.

The structure of monoclonal antibodies can be truncated in a variety of ways and yet maintain several of their functions, such as antigen (target) binding, antigen specificity, and cell killing. Four such molecules appear in Figure 5.8. An antigen-binding fragment (Fab) of a mab is displayed in Figure 5.8, indicating a structure containing two parallel linked chains: V_H (variable heavy) plus C_H (constant heavy), and V_L (variable light) plus C_L

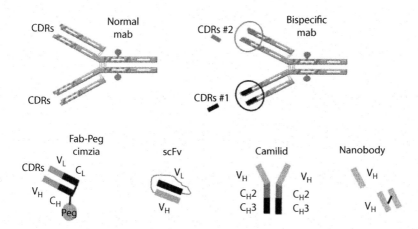

Figure 5.8 Monoclonal antibody engineering.

Abbreviations: CDR – complementarity determining region, C_H – constant heavy chain, C_L – constant light chain, Fab – antigen-binding fragment, mab – monoclonal antibody, PEG – polyethylene glycol, scFv – single chain variable fragment, V_H – variable heavy chain, V_L – variable light chain. (Reprinted with permission from Amgen Inc.)

(constant light), along with the CDRs maintained. This particular molecule is also pegylated to create the product, certolizumab (Cimzia). A variable double chain is presented (scFv), containing two variable domains, one light and one heavy. Camilid animals (camels and llamas) have monoclonal antibodies with simpler structures (only six heavy and no light domains), as shown in Figure 5.8. Nanobodies are very small antibody fragments consisting of only one or two variable heavy domains. Mabs also are engineered as antibody fragments to be combined with other proteins, creating fusion proteins with their combined actions, to be discussed below.

Molecular fusion

The concept with fusion molecules is to structurally combine two compounds, a mix of proteins, peptides, a mab fragment (Fab or nanobody), toxins, vaccines, or a chemical. Twenty-nine such molecules were commercially available as of the end of 2019. Goals have been to alter pharmacokinetics, result in combined properties for the full molecule, serve as a carrier to the site of action, or enhance absorption of the large biological molecules. Several examples have been achieved and marketed, for example, denileukin, a protein (Interleukin-2) and a protein toxin, diphtheria proteins, for renal cell carcinoma; abatacept, the CTL4-A protein extracellular domain (T-cell antigen protein) and a fragment of IgG1 for rheumatoid arthritis; the protein albumin bound to a drug, paclitaxel, for breast cancer; Pneumococcal vaccine, a conjugate of the polysaccharide of *Streptococcus pneumoniae* bound to diphtheria protein, preventing Pneumococcal pneumonia; and etanercept, combining IgG1 Fc fragment with tumor necrosis factor, a mediator responsible for inflammation in arthritides, colitis, and psoriasis. Table 5.3 further elaborates on the types of fusion proteins and examples.

A unique fusion protein involves a "peptibody" construction of an Fc fragment of IgG1 with 2 thrombopoietic peptides (14 amino acids each) in tandem connected to each arm of the IgG1 Fc molecule, romiplostim (Nplate®), to stimulate platelet production in thrombocytopenia. Other engineering of monoclonal antibodies includes glycosylation changes, such the molecule, obinotuzumab (Gazyva®), with reduced fucose CHO which is used for leukemias and lymphomas.

Table 5.3 Fusion proteins – types and examples

Antibody fragments + proteins:
- Antibody fragment (Fc domains of IgG1) + protein from disease receptor targets (usually extracellular domains) – four such proteins as examples:
  - Tumor necrosis factor (TNF) for etanercept (Enbrel®) in arthritis, interleukin-1 (IL-1) receptor for rilonacept (Arcalyst®) in Cryoprin associated periodic syndrome, CTLA-4 protein antigen for belatacept (Nulojix®) in organ rejection, vascular endothelial growth factor (VEGF) receptor (antigen) for aflibercept® in ophthalmology and oncology
- Antibody fragment, Fc of IgG4 + GLP-1 receptor antagonist in diabetes mellitus, plus changes in amino acid sequence, dulaglutide (Trulicity®)
- Antibody fragment, Fc of IgG1 + coagulation proteins, Factor 8, (Eloctate®) (B domain deleted also) or Factor 9, (Alprolix®) (B domain deleted also)
- Antibody fragment (Fc IgG1) + an enzyme's two polypeptide domains of catalytic site of tissue alkaline phosphatase + bone targeting domain, asfotase alfa (Strensiq®)

Interleukin directed proteins:
- CD-123 directed IL-3 + truncated diphtheria toxin, tagraxofusp-erzs (Elronzis®), in genetic enzyme disease
- IL-2 + diphtheria fragments of A & B toxins, denileukin diftitox (Ontak®), in cutaneous lymphoma cancer

Vaccines:
- *Neisseria meningiditis* polysaccharide + diphtheria toxoid, Menactra®, or diphtheria protein CRM197, Menveo®, for meningitis prevention
- *Streptococcus pneumoniae* saccharides + diphtheria toxin protein CDM 197, (Prevnar 13®), for Pneumococcal pneumonia prevention

Hormone:
- Glucagon-like peptide-1 (GLP-1) receptor agonist peptide + albumin, Semaglutide (Ozempic®), in diabetes mellitus

Biosimilars

The creation of "copycat" biological products, called biosimilars, currently presents many challenges to manufacture proteins by recombinant DNA technology and to "match" their structures and activities as close as possible in the laboratory and sustain their pharmacologic properties in patients versus a reference to U.S. Food and Drug Administration (FDA)-approved product. The basic tenet to this date by drug regulators in the United States and the European Union (EU) is "the same manufacturing process of recombinant DNA for biologicals at different companies or at different locales within the same company can NOT fully be the same process with the same outcome (product)." The reasons for biosimilars to not be the same at different companies are manifold, because of the complex structures of the protein molecules in question and the sophisticated and highly variable manufacturing situation employing living systems, as listed in Table 5.4. Biosimilars is the name used for these copycat protein and polypeptide products since a generic drug (identical chemical) is not possible. "Follow-On" products is another name used for biosimilars. The term "biobetters" is used to identify very similar biotech products structurally but possessing a superior biologic profile in the laboratory and in patients. The FDA also names biosimilars with the same proprietary name along with a four-letter suffix, for example, original product – bevacizumab (Avastin®), biosimilars – bevacizumab-awwh (Mvasi®), and bevacizumab-bvzr (Zirabev®).

Table 5.4 Complexity of protein and the protein manufacturing process

Protein issues	Manufacturing complexity
• Large very complex structures • Heterogeneity in each batch (isoforms) • Post-translational modifications manifold • MOA complex – antigen binding • Bioassay – complex, novel, & very manifold • Relatively unstable molecules • Immunogenicity potentially significant • Formulation variability • Adverse events unexpected	• Production in genetically modified organisms (GMOs) (living cells) • Host cells variability – type & growth • Vectors – types, transfection, & activity • Gene sequencing & insertion site vary • Gene add-ons: promoters & enhancers • Growth media & culture variables • Harvesting biologicals & cells – vary • No access to reference product data • Need comparison to reference product • Patents of molecule & processes

The FDA and EU guidelines promulgated in 2004 to 2008 for the UE and first in 2012 and later for the United States are detailed and extensive. Extensive biological testing is always required, in comparison to a reference product already approved by a regulatory agency. However, access is not permitted by the biosimilar manufacturer to the proprietary data and information held by the original company and regulatory agency. The application for approval for use and commercialization for a biosimilar is less than a new product approval, primarily for fewer Phase 3 clinical trials. Quality testing is extensive and involves the expression system (host cells, vectors, media, growth, harvesting), manufacturing process, structural analyses (amino acid sequence, disulfide bridges, peptide domains, glycosylation sites and composition, isoforms, folding), physicochemical properties, receptor binding, immunogenicity, functional analyses, impurities, stability, formulation. Pharmacological and clinical evaluations include animal data (pharmacodynamics, pharmacokinetics, immunogenicity, and toxicology), mechanism of action tests (targets, binding, signaling outcomes), clinical trials (head-to-head trials with the same end points, dosing schemes, and patient populations), clinical immunogenicity, safety, plus post-marketing surveillance after marketing. Generally, Phase 2 and not Phase 3 trials are required for product approval. The goal is very close similarity, but differences are acceptable if they do not impact the activity and clinical usage; however, any significant difference must be elucidated to the regulatory agencies and explained why it is not important in the action and clinical impacts of the product. A biosimilar can be studied in only one or two specific indications, yet can be approved for use in further indications found in the labeling of the original reference product as long as (1) the mechanisms of action are similar, (2) the dosing schemes are not different, (3) biologic and biochemical tests establish similarity, and it is without any indication of significant differences in (4) expected patient improvement or (5) adverse events. Also, a biosimilar product is not considered therapeutically interchangeable by regulatory authorities until additional information is available with biosimilar changeover studies between the original product and the biosimilar and no differences are established.

Biological product delivery and formulations

Most biological products (75%) are proteins, which, as described earlier, are large, complex, and relatively delicate molecules. Formulations of proteins are influenced by several protein traits, such as peptide lability, the three-dimensional structure, the charge on the protein, and their stability. Breakdown of proteins can occur through many mechanisms (chemical, physical, structural), as outlined earlier in Chapter 2 in Table 2.5. Further contamination and impurities are significant potential problems, given the living systems employed in their manufacture. The final formulations often are sensitive to temperature extremes, which can cause aggregation or precipitation. Restrictions in diluents exist because of potential adverse changes in stability. For example, filgrastim growth factor requires dextrose 5% in water and not saline for the diluent for molecular stability, and sargramostim growth factor, a very similar molecule, is the opposite. The protein formulations usually do not contain preservatives because of interactions of preservative compounds with proteins often causing instability of the protein, which necessitates single-use vials. Refrigeration is the norm to achieve maximal shelf life. Prior to administration, vials should be warmed to room temperature to lessen local reactions. Some proteins will require lyophilization (a dry frozen powder form instead of liquid) to create a practical longer shelf life duration. Excessive agitation is yet another possible denaturing problem for proteins, requiring care in the reconstitution of a lyophilized product. In creating the marketed product, the formulation for any biological molecule should avoid changes in activity of the molecule, its pharmacokinetics, its immunogenicity, its toxicity, and local irritation upon administration.

Also, specialized biotechnology-related molecular manipulation and formulations have been employed to enhance product delivery, which has been done for liposomes and polymers. Liposomes are biological molecules comprised usually of layers of lipids with a drug or biological incorporated into the center or a specific layer in the liposome structure. The goal is improved drug delivery to and activity in tissues because of the carrier function and lipid nature of the liposome, which in association with the lipid cell walls of the target tissue fosters more penetration into cells. Pegylation of the liposomal molecule is often used as well, to reduce potential immunogenicity of the product. Hopefully, adverse events from off-target tissue activity can be reduced through liposomes or pegylation. Polymers serve as carriers of a drug or biological to enhance product delivery to the tissue site, for example, a carmustine-polymer product allows placement of a cancer drug in the brain post-surgery.

chapter six

The science of biotechnology
Oligonucleotides and genetics

In drug discovery and development, genetic materials (DNA, mRNA, genes, and RNA enzymes) have become available in research as potential biological products and as primary tools of product discovery. Ten primary areas are covered here: polymerase chain reactions (PCRs), RNA inhibitory technologies (antisense, micro-inhibitory RNA, ribozymes), gene therapy, gene editing, aptamers, biomarkers, transgenic animals, and pharmacogenomics.

Polymerase chain reaction

PCR is a critical core process in biotechnology that permits substantial expansion of the amount of genetic material (DNA, genes), starting from minute amounts. As discussed in Chapter 2, first, the process involves denaturing DNA with high heat (90°C), that is, unraveling the DNA double helix so that the genetic code (sequence) can be read and possibly duplicated. Second, a leader sequence for DNA is used to bind to the target DNA sequence and initiates reading of the genetic code at a specific point in DNA. Both helices (strands) of DNA can be read, that is, duplication of the target DNA sequence. Third, the heat-stable enzyme from the bacteria *Thermus aquaticus*, a DNA polymerase, catalyzes the reading of the genetic code with incorporation of the four nucleotides and extension of the replicated DNA sequence. The four nucleic acids are provided as sources for DNA duplication (adenine, thymine, guanine, cytosine). By sequential repetition of these three steps, the genetic material is magnified; for example, 20 replications of the three steps yield a million-fold increase in the DNA material.

Oligonucleotide (RNA) therapeutics

Abnormal proteins can be responsible for a disease, and we can determine which cell and which abnormal mRNA produces the particular abnormal protein or absence of the protein. A genetic mutation leads to a mutated abnormal mRNA. Antisense involves synthetic oligonucleotides, often

RNA derivatives that bind to the aberrant mRNA, which alters the expression of genes and mitigate disease. In the translation process and before transcription, therapeutic oligonucleotides (anti-RNA molecules) can disrupt the mRNA function and prevent aberrant proteins (responsible for disease) from being produced in cells. Their mechanisms of action (MOAs) are at least five-fold: (1) altering the splicing of mRNA from DNA during translation, eliminating the abnormal mRNA; (2) a double-stranded anti-RNA therapeutic associating with the requisite Argonaute 2 protein in the intracellular RISC (RNA-inducing silencing complex), causing RNA cleavage; (3) a complementary microRNA (miRNA) directly binding to mRNA and blocking its activity; (4) recruiting enzymatic RNA (a ribozyme), RNase H, to cleave RNA with cessation of translation; and (5) oligonucleotide aptamers binding to proteins arresting their physiologic function. See Figure 6.1 for the diagrammatic presentation of three MOAs of antisense; A MOA - splicing change in mRNA, B MOA - RNA cleavage through RISC system, C MOA - inhibitory mRNA binding.

RNA molecules are challenging molecules to develop as therapeutics agents. The significant challenges include four main possible issues: (1) limited delivery of the anti-RNA molecule into cells, (2) poor stability of the product related to enzymatic degradation, (3) limited possible release of the molecule from its delivery system, and (4) lack of robustness of the

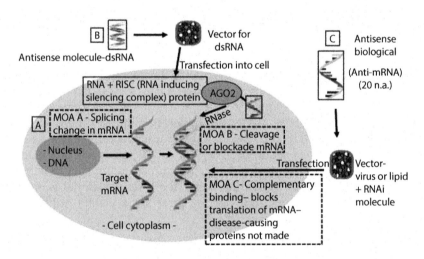

Figure 6.1 Oligonucleotide MOAs in mRNA function: (A) splicing disruption, (B) RNA cleavage through RISC system, and (C) inhibitory RNA binding to mRNA.

Abbreviations: AGO2 – Argonoute 2 protein chaperone, dsRNA – double-stranded RNA, MOA – mechanism of action, n.a. – nucleic acid, RNAi – RNA inhibitor, RNase – enzyme catalyzing RNA. (Reprinted with permission from Amgen Inc.)

antisense activity. Furthermore, the RNA molecules have several of the following specific functional limits that must be addressed for optimal pharmacologic activity: (1) degradation *in vivo* by many common blood and cellular nucleases; (2) transfection, that is, the need for incorporation into cells' cytoplasm or nucleus, their sites of action; (3) limited binding affinity to target RNA sequences; (4) moderate-sized molecules (10–30 times larger than small drug molecules) impacting transfection; (5) administration needed systemically and often more directly to locale of target cells (e.g., intra-ocular, intrathecal); and (6) very short duration of action. Structural modifications of the antisense RNA molecules improve their functionality; for example, (1) most common change of phosphorothioate alterations (replacing non-bridging phosphodiester oxygen with sulfur) – offering more stability versus nucleases, and more binding to serum proteins lengthening half-lives and longer exposure of RNA molecule to target cells; (2) ribose modifications such as replacing the 2'-hydroxyl group with 2'-*O*-methyl, or 2'-*O*-methoxyethyl, or 2'-fluoro groups, offering more stability and hybridization with better binding, leading also to more stability versus nucleases and more activity; (3) bridging between nucleic acids within the molecules, for example, bridge between the 2'-oxygen to the 4' carbon of ribose, offering better hybridization and possibly use of smaller-sized molecules; (4) attachment of small lipid carbon chains to nucleic acid chain for more stable molecules; (5) use of nanoparticle liposomes incorporating the nucleotide molecules, offering nuclease protection, intracell delivery, and longer serum times; and (6) conjugation of *N*-actylgalatosamine (GalNAC) to the nucleotide structure, offering attachment to GalNAc receptors on cells and improving internalization to intracellular sites of action. Such structural improvements have led to superior functioning molecules and several new product approvals over the prior five years.

The antisense RNA molecule will bind to the abnormal mRNA molecule (intracellular), preventing the translation process and the disease from manifesting. One product was approved as the proof of principle, fomivirsen (Vitravene®), and was used to treat cytomegalovirus (CMV)-associated retinitis that can occur in AIDS patients; however, its administration directly into the eye and, especially, improved HIV therapy with much fewer CMV infections limited its usefulness and led to marketing withdrawal. In addition, micro-inhibitory RNA molecules are now being developed to attack disease-producing RNA. Inhibitory RNA molecules possess excellent specificity and function to turn off aberrant RNA, but the many drug development challenges noted above exist along with the challenge of optimization of the miRNA molecule for efficacy. Six oligonucleotide anti-RNA molecules were developed from 2013 to 2019 based on improved molecules for five previously untreatable rare diseases such as (1) Duchenne's muscular dystrophy – eteplirsen (Exondys 52®), (2) spinal muscle atrophy – nusinersen (Spinraza®), (3) polyneuropathy in hereditary transthyretin amyloidosis – inotersen (Tegsedi®), (4) – patisiran (Onpattro®), and (5) hepatic veno-occlusive disease – defibrotide (Defitelio®).

Ribozymes are RNA molecules comprising sequences of nucleic acids that possess enzymatic catalytic properties and bind to specific sites in DNA or RNA and cleave the chain. A ribozyme will have subunits responsible for the binding function and subunits responsible for enzyme function. They generally have the following desirable traits: specificity in targeting, cleavage of target RNA, small size amenable to formulation and dosing, and multiple turnover (one molecule binds and acts and then moves on to next molecule and repeats its function). However, challenges are manifold; for example, the need for cell insertion (transfection), nuclease protection in the blood, and chaperone proteins for movement in cell cytoplasm. No products have yet been approved. Aptamers are small oligonucleotide molecules that bind to proteins to disrupt disease pathogenesis. Their benefits are small molecular size, specificity toward target protein, and low immunogenicity; however, limitations are nuclease susceptibility, short systemic half-lives, and possibly limited affinity to targets. Proof of principle has been achieved as one aptamer has been approved for use, a pegylated conjugate of an oligonucleotide for wet acute macular degeneration, pegaptanib (Macugen®). The pegylation protects the molecule from lysis from nucleases, and offers a longer serum half-life and a measure of immunity.

Gene therapy

Gene therapy is a technology employing a gene (a DNA sequence responsible for producing a protein) as a therapeutic agent to treat a disease caused by an abnormal or nonfunctional gene. The gene's major benefit as a therapy is its curative outcome with only a single administration. The potential mechanisms and goals of gene therapy include five actions: (1) replacement of an inactive gene; (2) reactivating inactive genes; (3) turning off genes causing disease, as in oncogenes; (4) turning on further naturally protective genes; or (5) adding a gene characteristic to cells, such as increased susceptibility of cancer cells to chemotherapy drugs. There are substantial development challenges in gene therapy to creating a viable product, some of which follow: (1) finding the target gene that is causing the disease or needing enhancements, (2) identifying the optimal target human cell for the disease and patient for insertion of an additional gene, (3) inserting (delivery) an extra gene reliably into human cells *ex vivo* or *in vivo*, (4) ensuring the gene payload maintains stability and activity as it traverses the physiologic processes (e.g., nuclease enzyme attack), and (5) causing the gene to be functional in a reasonably physiologic manner (extent and duration of activity). Figure 6.2 offers a representative simplistic

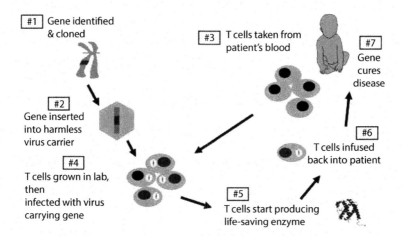

Figure 6.2 Gene therapy – POP in SCID (1990).

Abbreviations: POP – proof of principle, SCID – severe compromised immune disease. (Reprinted with permission from Amgen Inc.)

generalized illustration of gene therapy, which first was created in 1990 as proof of principle for gene therapy for severe compromised immune disease (SCID). The seven steps of gene therapy process are elucidated.

Although science is rapidly identifying many gene mutations in the human genome responsible for disease, gene delivery achievement in a reliable, reproducible manner remains a most daunting challenge. Viruses are most commonly used for a gene delivery vector due to their natural ability to carry genetic material and deliver it into human cells (infect or transfect), and allow the genes to be turned on. Five types of viruses as vectors are being studied: retrovirus, lentivirus, herpes virus, adenovirus, and adeno-associated virus; in addition, plasmids can be used as well. The optimal gene vector has a lengthy and daunting list of optimal traits desired for its efficacy, as listed in Table 6.1.

Additionally, clinical use of a gene/vector product needs ease of administration with practical cell/genetic manipulation and safety in patients. Manufacturing needs for a vector/gene package include ease of fabrication, inexpensive synthesis, and facile purification. These manifold challenges have been overcome successfully with the regulatory approval worldwide in 2016–2019 of four gene therapies: alipogene tiparvovec (Glybera®) for lipoprotein deficiency (adeno-associated virus serotype 1 vector), autologous gammaretroviral gene therapy (Strimvelis®) for SCID (CD34+ cells with gamma-retro virus vector), both in Europe, plus voretigene neparvovec (Luxtruna®) for retinal atrophy (adeno-associated virus type 2 vector) and onasemnogene abeparvovec (Zolgensma®) for spinal muscular atrophy (vector – recombinant adeno-associated virus 9).

Table 6.1 Exemplary traits or issues of virus vectors in gene therapy

- DNA versus RNA virus
- Protection of DNA package *in vitro* in vials
- Targeting to a specific cell type
- Endolysosomal protection in cells
- Cytoplasmic transport with chaperones
- Efficient unpackaging in nucleus
- Gene activation (transcription & translation)
- Robustness of gene expression
- Internalization of vector/gene by cells (transfection or endolysosome incorporation)

- Packaging of genes of varied sizes
- Serum stability with nucleases
- Infection of nondividing target cells
- Endolysosomal escape of gene
- Nuclear localization and transport
- Integration of gene into genome
- Stability of the gene expression
- Duration of gene expression
- Preexisting immunity absent
- Adverse effects limited (especially mutagenesis and inflammation)

Pharmacogenomics

Pharmacogenomics is the study of genetic phenotypes of patient subgroups in a disease and their genetic impact on drug actions, changing drug pharmacokinetics and/or activity (more or less pharmacologic effects). This area is also called "personalized medicine" wherein an individual's genetic makeup can predict efficacy or safety of a product. Over 60,000 single nucleotide polymorphisms (SNPs) exist in exons of genes that are genetic mutations, deletions, insertions, or repeats. For drug activity, these SNPs can have several impacts: They can change (1) the action of liver metabolic enzymes (and bioavailability), (2) receptor activity, and (3) drug transport. These mechanisms can lead to more or less adverse events, dosing changes up or down to achieve the same effect, more or less efficacy, and disease subtypes with different drug responses. Figure 6.3 presents two fictional SNPs and their impact on dosing and drug activity. The left frame gives two outcomes for bioavailability, amount of drug in the blood, because of SNP #1 in liver enzymes changing metabolism, or the SNP in drug transport with differing drug passage across the mucosa. SNP #2 displays different receptor sensitivity over time with drug concentrations. A low drug concentration at high receptor sensitivity value (very sensitive receptor) offers high efficacy at low doses. In the right frame, the bioavailability concentrations are superimposed on the graph with the receptor sensitivities and demonstrate complexity in the combined SNP effects, depending on dose, bioavailability, and receptor activity. Also, the toxicity drug concentration line is superimposed as well; the high bioavailability SNP exceeds the toxicity threshold with more side effects being the result.

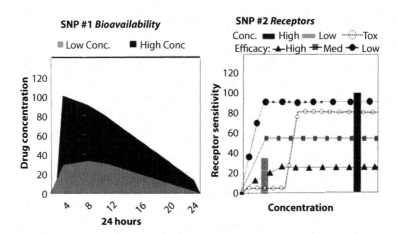

Figure 6.3 Pharmacogenomics: SNPs impact on bioavailability and receptor activity.

Abbreviations: Conc – concentration, Med – medium, SNP – single nucleotide polymorphism, Tox – toxicity. (Reprinted with permission from Amgen Inc.)

Hopefully, we can identify (diagnose) these abnormal, or just different, genetic phenotypes responsible for different disease presentation or drug actions, and either use a drug in a more effective subpopulation or change dosing for less side effects or more drug efficacy. For example, in oncology, some breast cancer patients will have the genetic abnormality, Her2Neu gene-positive (an oncogene), leading to a more aggressive and fatal disease course. Fortunately, a mab has been developed against this genetic subpopulation, trastuzumab (Herceptin), turning off the oncogenic mutations and offering more efficacy, but only in this subgroup of patients. Viability of pharmacogenomics in patient care requires diagnostic tests to be identified, validated, and commercialized to identify these patients at risk, and then requires the studies to establish appropriate drug doses for the unique phenotypes or drugs that will work in these unique phenotypes. Advantages to the health care system include higher drug efficacy in a disease subgroup, less use of drugs where they will not work, and, possibly, avoidance of drug toxicity.

Biomarkers

The diagnosis of disease basically has always been dependent on various disease indicators and tests related to normal human physiology and the pathology of the disease – such as liver enzymes, the kidney's urine electrolytes and proteins, and cardiac enzymes – in order to judge organ function or their disease status. These diagnostic factors are essentially disease biomarkers. Genetics has revealed added genetic mutations as biomarkers in identifying patient-specific differences in disease occurrence, pathogenesis, or severity. In addition, pharmacogenomics takes genetics one step further in associating differences in drug activity in definable subpopulations of patients with specific genetic differences (inherited or acquired), especially drug metabolism. A drug or biological product that is active in this subpopulation with a genetic variation may only be effective in that subpopulation of patients, requiring a biological disease indicator, a biomarker, to identify these different patients. Thus, another outgrowth of biotechnology is the biology of disease and the search for biomarkers to identify groups of patients with different responses to certain drugs, either drugs or biologicals. Therapy thus can be better individualized and given to the patient most likely to respond or used to avoid a predictable adverse event related to genetic differences. Several biological products are now effective and are marketed for a certain disease and a specific subpopulation, based on a biomarker. The trastuzumab example noted above was for breast cancer. The PK inhibitor drug, imatinib, is only effective in chronic myelogenous leukemia with Philadelphia chromosome-positive status. The mab proteins, cetuximab and panitumumab, are active usually only in patients with EGFR-positive and KRAS mutation-negative colorectal cancer cells. These three exemplary treatments require a diagnostic test for the respective genetic biomarkers to document the presence or absence of the genetic abnormality, engendering a higher likelihood of drug response. Biomarkers have several prerequisites: They (1) need to relate directly to the disease pathogenesis and the genetic or physiologic abnormality, (2) need to be consistently present over the course of the disease, (3) need to possess reliability to avoid false positives or negatives, (4) must be validated in clinical trials, (5) need to be reasonably practical in their conduct for anyone skilled in lab procedures to perform, and (6) must not be too expensive. Product development for a drug or biological with such a pharmacogenomic biomarker requires a dual development process and approval for both the product and the genetic test system.

Transgenic animals

Transgenic animals are a special genetics-related development in biotechnology, as well as all drug development that enhances the screening of potential therapeutic molecules. Through genetic engineering, an animal's genetic makeup (often in mice and rats) is altered by knocking out the animal's normal genes and inputting a human gene that can produce a target disease, which results in the animal presenting with a disease that is more human-like in its pathology, essentially a human disease model in a rat. A potential new drug candidate is administered to these transgenic animals, and the animal's disease model will better predict (respond to) the drug's action as being representative of what would really occur in humans.

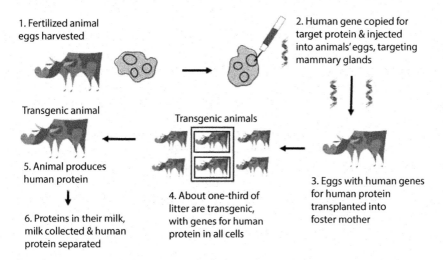

1. Fertilized animal eggs harvested

2. Human gene copied for target protein & injected into animals' eggs, targeting mammary glands

Transgenic animal

Transgenic animals

5. Animal produces human protein

6. Proteins in their milk, milk collected & human protein separated

4. About one-third of litter are transgenic, with genes for human protein in all cells

3. Eggs with human genes for human protein transplanted into foster mother

Figure 6.4 Transgenic animals for protein production. (Reprinted with permission from Amgen Inc.)

Another use of transgenic animals is for the manufacture of biological products actually in the animal (Figure 6.4). Through genetic engineering, human genes are inserted into the eggs of an animal, incorporating a human gene into the genome of the animal and expanding its genetic makeup. The foster mother with some eggs containing human genes will give birth to a litter of offspring, some of which will be transgenic with the additional feature of producing a human protein. A herd of transgenic animals then can be bred. Gene insertion often engages the mammary structure of, usually, sheep, goats, rabbits, or cattle, such that the human gene produces a human protein in the animal's milk, which can be harvested and separated. Products as of 2019 are commercially available from transgenic animals, for example, Atryn®, anti-thrombin protein, in goat's milk, and Ruconest®, conestal alfa, in rabbit's milk (for hereditary angioedema).

Genomics

The study of genomics in simple terms involves the identification of the sequence (genotype) and activities of all the approximate 25,000–30,000 genes in the human body, and their expression into proteins in specific patient groups (phenotype). Genomics is a technology in which we search for a gene that is responsible for some process or product in the human body. The gene may lead to a new process or disease target for which drugs or biologicals could be developed to favorably alter. For example, the gene for cell immortality was discovered, which led to the protein telomerase, an enzyme. Telomerase is responsible for adding telomeres, short nucleic acid sequences, to the end of all chromosomes, protecting chromosomes from mutations and, ultimately, cell death. Alternatively, genomics may lead to a gene that produces a protein that can be used as a therapeutic agent. For example, the protein osteoprotogerin is the natural substance that turns off osteoclasts in bones, the cells that break down bones, leading to osteoporosis. Genomics requires the collection of massive amounts of information regarding the human genome or genetics, including protein and peptides, receptors for activity of proteins or drugs, mechanisms of drug activity, and subunits of drugs, proteins, RNA, or DNA responsible for physiologic or pharmacologic actions. The science of bioinformatics now exists to store, share, integrate, analyze, and manipulate these massive amounts of data through facilitated methods using computers and algorithms to identify new products.

Gene editing

In specific chromosomes, gene editing achieves changing DNA sequences (genes) through gene deletions, mRNA changes, or posttranslational modifications, which alter gene expression, which changes protein production or functions. The methods of gene editing involve six steps: (1) entry of genetic tool into cells and their nucleus, (2) identifying and targeting specific DNA sequences, (3) DNA cleavage, (4) DNA removal and/or insertion, (5) follow-up changes in gene transcription, and (6) change in protein products and actions (new, altered, suppressed, or overexpressed). The early principle discovery that permits genetic manipulation was the discovery of restriction endonucleases that permitted DNA cleavage.

In this short overview, five methods for gene editing will be defined and briefly outlined: (1) ARCUT, (2) meganucleases, (3) ZFN, (4) TALEN, and (5) CRISPER/Cas. ARCUT is artificial restriction DNA cutter. The DNA cleavage involves a pseudo-complementary peptide nucleic acid that specifies the cleavage site, DNA excision and splicing with ethylenediaminetetraacetic acid and cerium, and DNA ligase to foster DNA attachment at the target site. Meganucleases are large protein enzymes that are many in number and naturally occurring and that excise DNA sequences. They are bound to proteins that assist in specifying DNA cleavage sites. They are limited by also naturally occurring repair processes in cells that can also cause changes in other DNA sites. Zinc finger nucleases (ZFNs) are synthetic programmable combinations of a restriction endonuclease (FokI) and small zinc-ion regulated binding domain proteins, which target triple codons (three nucleic acid sites). FokI nucleases are the DNA cleavage domain only with deletion of the DNA recognition domain. FokI requires homodimerization at the target site in order to cleave DNA, such that two zinc finger molecules are needed to target two nearby DNA sites for DNA cleavage. TALEN is a transcription activator-like effector nuclease, a synthetic construction of a restriction endonuclease (FokI also), bound to a DNA-binding protein domain (TAL effector). The TALEN can bind to single nucleic acids and functions similar to the ZFNs.

CRISPER/Cas9 is a clustered regulatory interspaced short palindromic repeat DNA sequence along with a CRISPER-associated protein enzyme with nuclease function. The Cas9 in CRISPER is formed from two short RNA molecules: a guidance RNA (gRNA) with a transactivating (tracrRNA) complex, which cleaves the DNA. Furthermore, another molecule involved in order for CRISPER to work is the protospacer adjacent motif (PAM), which needs to be adjacent to the acquired spacer sequence. Then, a desired DNA sequence can be inserted. In comparing gene editing methods, a challenging list of characteristics is desirable as follows: simplicity in design, engineering feasibility, multiple genome editing, specificity in targeting, efficiency in operations, reasonable cost, minimal off-target effects, minimal immune reactions, and no cytotoxicity. The uses are quite manifold: diagnostic utility, clinical usage, gene therapy, possible epigenetic utility, possible gene knock-out activity, RNA editing possibility, and mitochondrial DNA activity. The science and applications of CRIRPER/Cas9 gene editing is advancing rapidly creating new exciting treatment modalities with broad applications.

chapter seven

Laws and regulations governing biological product approval and usage

Biotechnology is a heavily regulated technology and industry by government agencies and health care systems with many laws and regulations. Biological laws and regulations have been promulgated over 140 years from 1883 to 2019 to protect the health and well-being of society and ensure the highest scientific and ethical standards in the research, manufacture, usage (efficacy and safety) of products, and business practices.

Figure 7.1 displays the myriad of government agencies, scientific professional groups, and general public entities (#15+) that ALL oversee and impact the biopharma industry and all related facets of operations and health care. This chapter will focus first on the laws and regulations from the U.S. government and, especially, the U.S. Food and Drug Administration (FDA) that control and guide the industry and the use of drugs and, especially, biologicals. Bolder print in Figure 7.1 for the arrows suggests more control or impact.

The influence of many agencies and groups on the conduct of drug and biological research, manufacturing, and product usage will be listed briefly for completeness, fleshing out the groups identified in Figure 7.1. After this brief discussion, the focus will be on the FDA, which has the most impact on the biotech and pharmaceutical industries. In product development and especially product usage, the overall basic questions are: How and when will a product become available for patient care use by health care professionals? Which product is to be used? For which patient? How-When-Where-By whom will it be used? For how much cost? Several government agencies besides the FDA directly influence biotechnology and biotech products as follows. Congress authors the new laws governing at least health care, industry practices, product use, compensations for them, and government agency policies and actions of oversight and control. The U.S. Department of Agriculture (USDA) creates guidelines in animal use in research of drugs and biologicals. The Center for Medicare and Medicaid Services (CMS) creates policies that dictate product usage and actually makes payments for products. The Veterans Administration (VA) and Department of Defense (DOD) both can also set

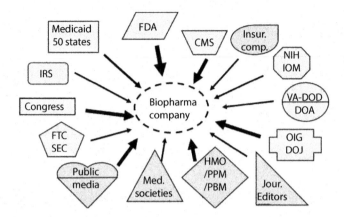

Figure 7.1 Biopharma industry – under a microscope in the United States and the rest of the world.

Abbreviations: CMS – Centers for Medicare and Medicaid Services, DOA – Department of Agriculture, DOD – Department of Defense, DOJ – Department of Justice, FTC – Federal Trade Commission, HMO – health maintenance organizations, Insur. Comp. – Insurance companies, IOM – Institutes of Medicine, IRS – Internal Revenue Service, Jour. – Journal, Med. Societies – Medical Societies, NIH – National Institutes of Health, OIG – Office of Inspector General, PBM – prescription benefit managers, PPM – physician practice management groups, SEC – Security and Exchange Commission, VA – Veterans Administration.

policies for usage and make payments for veterans and military personnel, respectively. The legal oversight of biological product usage, all operations of the industry, and proper follow-through on all laws is conducted by the Office of Inspector General and Department of Justice. Business practices in sales and marketing, use of tax credits in research, or other financial incentives with health care institutions are overseen by the Internal Revenue Service (IRS). The Federal Trade Commission (FTC) and Security and Exchange Commission (SEC) monitor the financial reporting, contracts with nonindustry groups, public financial offerings, and related necessary science reports and announcements being factual and not misleading to the public. State Medicaid agencies (#50) generally follow the CMS and implement the programs established under the CMS, but they can also create added regulations. Therefore, the states' impact on industry research and marketing needs to be consistent with federal programs, but they can vary and complicate industry practices. The delivery of health care is often organized through hospital systems (public and private), health maintenance organizations (HMOs), physician practice management groups (PPMs), and prescription benefit managers (PBMs), all of which set guidelines for products' usage and actually purchase or pay for products. Insurance companies pay for health care and product usage, including writing usage guidelines for health care professionals and the public. Science and medical groups in government and independent private organizations promulgate guidelines for the research of products (design of trials for example) and treatment guidelines for all disease areas, for example; the National Institutes of Health (NIH), Institute of Medicine (IOM), Centers for Disease Control and Prevention (CDC), and medical societies (for all the many medical disciplines). Even medical journal editors influence industry for biologicals and drugs by setting drug design guidelines used in clinical studies as being acceptable for consideration of publication of the study results and hence distribution of data and findings for medical professionals, the public and media, and regulatory agencies. Finally, the media and general public monitor and evaluate the biopharma industry and health care and offer opinions about all health care issues, drug and biological approvals, availability and usage, the actions of government agencies (or lack thereof), and cost of and payment for biological and drugs. They can exert pressure on all the other agencies and groups and basically have a multiplier impact for their comments and actions.

Table 7.1 enumerates the extensive list of laws and regulations, along with the primary action created, and a designation if biologics (*B*) or drugs (*D*) were involved in the laws. In the first 60 years in the twentieth century, the new drug and biological laws primarily resulted from major health care disasters and public outcries for regulation; for example, tetanus exposure due to contamination of diphtheria anti-toxin (1900s) – 1906 law; sulfanilamide elixir contamination with diethyl glycol and deaths – 1938 law; thalidomide sedative causing birth deformities – 1962 law. Concerns about exploding drug costs led to laws fostering less costly product alternatives, in 1984 – generic drugs facilitation and 2010 – biosimilar biological products' approvals. Desire to facilitate drug research and foster faster approvals to expedite product use for improved outcomes in patients with serious and life-threatening diseases resulted in several laws to offer advances in patient care sooner than later; for example, in 1983 (orphan diseases treatments); 1992 (serious and life-threatening diseases – priority reviews and surrogate disease markers); 1997 (serious and life-threatening diseases – fast track reviews); 2002 and 2003 (Research in pediatric diseases); 2007 (neglected and rare disease treatments); 2012 (serious and life-threatening diseases – breakthrough therapies); 2017 (tissue/cell therapies expedited – serious diseases); and 2018 (oncology drugs specifically expedited).

Besides laws and regulations for drugs and biologicals permitting FDA oversight and actions for the protection of the health and welfare of the public, the FDA promulgates written "guidances" for the biopharma industry practices in areas of the research, clinical development, manufacturing, marketing and sales. The guidances assist the industry in setting

Table 7.1 U.S. government drug laws and regulations

1. 1883: *B/D*. Division of Chemistry in U.S. Department of Agriculture (USDA)	Created for food research
2. 1902: *B*. Biologics Control Act	Safety-Purity-Potency focus of biologics
3. 1906: *B/D*. Pure Food & Drug Act	Adulteration of products prohibition
4. 1907: *B/D*. Board of Food and Drug Inspection in USDA	Importance of inspection recognized
5. 1908: *B/D*. Referee Board of Consulting Scientific Experts in USDA	Importance of science basis in regulatory decisions recognized
6. 1912: *B/D*. Shirley Amendment	Outlaw of fake medical claims
7. 1927: *B/D*. Food, Drug & Insecticide organization	Created to focus regulatory work in USDA
8. 1930: *B/D*. Food, Drug and Insecticide Administration	Name change to FDA
9. 1938: *B/D*. Food Drug & Cosmetics Act	Safety requirements in drug approvals
10. 1944: *B*. Public Health Services Act	Biological product licensing created
11. 1951: *B/D*. Durham-Humphrey Amendment	Prescriptions required, written by medical professionals
12. 1962: *B/D*. Kefauver-Harris Amendments	Efficacy requirements in product licensing and approval for use Inspection of manufacturing facilities Advertising only for approved uses
13. 1976: *B*. Bureau of Biologics	Created in FDA & moved from the National Institutes of Health
14. 1983: *B/D*. Orphan Drug Act	Orphan drug designations for rare diseases Incentives – financial [clinical grants, Prescription Drug User Fee Act (PDUFA) fee waiver, tax credits], data requirement lessened, patent exclusivity extended 7 years
15. 1984: *B/D*. Hatch-Waxman Act	Formal generic drug approval process Patents extended for original product licensee
16. 1992/1997/2002/2007/2012: *B/D* –PDUFA	Fees for applications from companies Infusion of staff for applications' reviews Priority reviews, 6 vs. 10+ months Accelerate review, surrogate end points FDA & sponsor meetings encouraged Special protocol assessments from FDA Sunshine law, renewal every 5 years

(Continued)

Table 7.1 (Continued) U.S. government drug laws and regulations

17. 1997: *B/D*. Food & Drug Modernization Act	Fast track designation
18. 2002: *B/D*. Best Pharmaceuticals for Children Act	Studies in children fostered but voluntary requests only Patent extension for 6 months
19. 2003: *B*. Biotech products from Center for Biologics Evaluation and Research (CBER) to Center for Drug Evaluation and Research (CDER)	CDER's Office of Biotechnology created for therapeutic proteins & monoclonal antibodies
20. 2003: *B/D*. Pediatric Research Equity Act	Research in children mandatory Waivers permitted if appropriate
21. 2007: *B/D*. Priority Review Voucher	Neglected/rare diseases accelerated reviews (faster with fewer patients in rare diseases) Transferable & sold to other sponsors
22. 2008: *B/D*. FDA Amendments	Adverse event oversight elevated
23. 2010: *B*. Biologics Price Competition & Innovations Act	Biosimilar licensure created
24. 2012: *B/D*. Breakthrough Therapy Designation	Accelerated product approval
25. 2017: *B*. Regenerative Medicine Advanced Therapy	Accelerated approvals for regenerative medicines, e.g., cells, tissue
26. 2018: *B*. Biosimilar User Fee Act	Biosimilar biotech product approvals process initiated and application fees described
27. 2018: *B/D*. Real-Time Oncology Review	Accelerated oncology drug approvals

standards to be addressed and followed and setting regulatory expectations in at least eight areas: (1) FDA definitions of key elements in the laws and regulations, (2) what is considered by the regulatory agency as best practices (required and suggested), (3) requirements for licensing and approval for general use for drugs and biologicals, (4) required and suggested processes and procedures in performance of research and clinical development, (5) required and suggested processes and procedures in manufacturing, (6) marketing and advertising activities under scrutiny, (7) product use guidelines by medical professionals and patients, and (8) requirements to establish product quality (e.g., stability, potency). Over 800 guidances exist for both drugs and biologicals, and biologicals have over 100 specifically for that area. Table 7.2 lists over 30 areas wherein guidances exist for biologicals, often with many guidances in any one area, for example, for biosimilar biotechnology products [new similar copies of

Table 7.2 Types of U.S. Food and Drug Administration (FDA) guidances impacting biologicals [100+ out of 800+]

1+. Advertising & Promotion, 2. Advisory committees to the FDA, 3+. Applications for approval [#9 types: abbreviated new drug application (ANDA) – generic drugs, biological license application (BLA) – biologics, investigational new drug (IND) – investigation in patients permitted, new drug application (NDA) – new drug, premarket approval (PMA)-premarketing [devices], prescription drug change to over-the-counter use, supplemental NDA (sNDA) – new indications or doses, 351(k) – devices, 510(k) – biosimilars, 4+. Biologicals (recombinant or synthetic or naturally occurring); categories include biosimilars, blood derivatives, cell therapies, gene therapy, monoclonal antibodies, peptides, proteins, ribonucleic acid inhibition(RNAi) therapy, tissue therapy, vaccines), 5+. Biostatistics, 6. Breakthrough therapy designations, 7+. Clinical development (data & design & procedures; Phase 1, Phase 2, Phase 3, postmarketing – Phase 4), 8. Common Technical Document, 9. Compounding, 10+. Biologic testing, 11+. Disease specific guidances, 12+. Good practices – clinical & laboratory & manufacturing and review practices (GCP, GLP, GMP, GRP, respectively), 13+. Import/export guides, 14+. International harmonization, 15. Labeling requirements, 16+. FDA letters (e.g., complete response letter, notice of violation, warnings), 17+. Manufacturing (product – chemistry-manufacturing-controls, formulation & stability, processes, facilities), 18. Medication guides [patient package insert (PPI)], 19. Orphan drugs, 20+. Pediatrics' research and usage, 21+. Pharmacology, pharmacokinetics, pharmacogenomics, 22. Priority review vouchers, 23+. Product licensures (biologicals – new & biosimilars; drugs – new & generic & over the counter; devices; combinations), 24+. Product specific, 25+. Quality [chemistry, manufacturing, controls (CMC); microbiology; analytical testing], 26. Rare diseases, 27. Regenerative medicine therapy, 28+. Research (preclinical), 29+. Safety (adverse events reporting, immunogenicity, postmarketing surveillance, pharmacovigilance, risk evaluation and mitigation systems), 30+. Toxicology (preclinical and clinical), 31+. User fees (new drugs, new biologicals, generic drugs, biosimilars, devices), 32. Meetings with the FDA by sponsors.

Note: + sign next to each type of guidance indicates multiple guidances exist within that category.

proteins and monoclonal antibodies mabs)]. In addition, Table 7.3 lists eight such specific guidances for biosimilar class of biotechnology products. All details for all these guidances can be found in the FDA.gov web site.

Biosimilar biotechnology products (mostly proteins and mabs) are copycat products very similar to and compared to a marketed reference product. However, they are not considered generic (exact duplicates, as with drugs), because their manufacture using living systems, like recombinant DNA processes, leads to differences in the molecules that may impact efficacy and safety. A basic tenet is the same manufacturing process by the same company at a different locale can NOT be the same, resulting in molecular differences. Any differences in structure and function must be identified and examined for their impact, if any, especially not significantly impacting safety, stability, or toxicity. The "totality of the evidence" is reviewed by FDA, and a streamlined regulatory application for approval is possible with extra emphasis on the chemistry and characterization of the new biosimilar molecule compared to the reference standard and less clinical trials (often Phases 1 and 2, but not necessarily Phase 3). A variety of guidances have been created for biosimilars (see Table 7.3). Patent protection for 12 years is given to the company's product.

The naming of biotechnology products for their nonproprietary names was revised in 2015 by the FDA for biosimilar products to improve differentiation between similar products. A four-letter suffix was to be used going forward, e.g., filgrastim-sndz for Zarxio by Novartis in 2015, etanercept-szzs for Erelzi by Novartis in 2016, and infliximab-dyyb for Inflectra by Pfizer in 2016. In January 2017, the FDA promulgated a new guidance, nonproprietary naming of biological products, involving the use of a four-letter suffix to be affixed at the end of the nonproprietary names for all new biotech products that are proteins, fusion proteins, monoclonal antibodies, and gene therapies, as well as biosimilars. The overarching goal is to enhance pharmacovigilance especially for adverse events with better differentiation between similar products. The four-letter suffix has six criteria,

Table 7.3 U.S. Food and Drug Administration (FDA) guidances
for biosimilar products

---

1. New and Revised Draft Q&As on Biosimilar Development and the Biologics Price Competition and Innovation (BPCI) Act (Revision 2), Draft Guidance for Industry. Center for Drug Evaluation and Research/Center for Biologics Evaluation and Research (CDER/CBER), December 2018
2. Questions and Answers on Biosimilar Development and the BPCI Act; Guidance for Industry. CDER/CBER, December 2018
3. Considerations in Demonstrating Interchangeability with a Reference Product; Draft Guidance for Industry. CDER/CBER, January 2017
4. Clinical Pharmacology Data to Support a Demonstration of Biosimilarity to a Reference Product; Guidance for Industry. CDER/CBER, December 2016
5. Labeling for Biosimilar Products; Guidance for Industry. CDER/CBER, March 2016
6. Scientific Considerations in Demonstrating Biosimilarity to a Reference Product; Guidance for Industry. CDER/CBER, April 2015
7. Quality Considerations in Demonstrating Biosimilarity of a Therapeutic Protein Product to a Reference Product; Guidance for Industry. CDER/CBER, April 2015
8. Reference Product Exclusivity for Biological Products Filed Under; Draft Guidance for Industry. CDER/CBER, August 2014

as follows: nonproprietary names, unique, lowercase letters, devoid of meaning, attached with a hyphen at the end of name, and free of legal barriers. The names of all biotech products before 2017 were not altered with a suffix.

For product approvals, a company must establish the triumvirate of efficacy, safety and quality for a drug, biological, or device in the applications for approval at the FDA and the European Medicines Agency (EMA). Applications include the investigational new drug (IND) application, allowing initial use in humans; for product approvals by the FDA, applications include the new drug application (NDA) for new *drugs*, biological license application (BLA) for *biological* licensing, premarket approval (PMA) for premarket approval of brand new *devices* or 351(k) for new similar devices; supplemental NDA (sNDA) or supplemental BLA (sBLA) for supplementary applications for new indications, new dosages, and new formulations; abbreviated new drug application (ANDA) for *generic* drugs; and 510(k) application for *biosimilar* biological products.

To accelerate approval of products and improve availability of therapies for patients, the federal government has empowered the FDA with a series of accelerated product approval processes. The focus has been on serious and life-threatening diseases, as long as significantly improved treatments are the goals. Table 7.4 provides seven laws initiating six new regulations, along with the qualifying criteria for the disease and drug or biological, and the features/benefits of accelerating the approval process. The laws are designated as "Fast Track," "Accelerated Approval," "Priority review," "Breakthrough therapy," "Infectious disease – qualified products," and "Orphan Drugs – rare diseases," "Regenerative medicine," and "Antibacterial and antifungal drugs."

Many other systems exist for FDA oversight of biopharma industry and drug usage by the medical-pharmaceutical community, such as, (1) *Communications in letters*, such as, complete response letters regarding product approvals or not, and warning letters to change behavior in advertising, study trials, or manufacturing, or any significant issue involving inappropriate operations; (2) *Supplemental approvals* for added indications, doses, dosage forms; (3) *Product recalls, Marketing withdrawals, and Clinical research holds*, usually related to major unexpected adverse events or significant new product quality defects; (4) *Labeling* of products (actual product label on bottles and boxes) drafted by companies but edited and approved by the FDA; (5) various *Inspections* and resultant *warnings* or action requirements, covering manufacturing, clinical trial sites, and company and medical documents' retention; (6) *Advertising* of products by companies (accuracy, fair balance, and truthfulness); (7) *Quality assurance* with product testing (active ingredients and other content; purity, stability, contamination); (8) *Adverse event* reporting requirements, including ad hoc reports and also risk evaluation and management systems; (9) *Export certificates* for active ingredients and products distribution; (10) *Off-label* indications and related *publication*

Table 7.4 Accelerated approval processes for product approvals with the U.S. Food and Drug Administration (FDA)

Issue	"Fast track"	"Accelerated approval"	"Priority review"	"Breakthrough therapy"
Law	FDAMA 1997	FDC Regs 1992	PDUFA 1992	FDASIA 2012
Qualifying criteria	• Serious conditions, e.g., HIV/AIDS, cancer • Unmet medical need potential	• Serious conditions • Meaningful advantage over existing treatment • **Surrogate or clinical end point allowed for efficacy**	• Serious conditions • Significant improvement in safety or efficacy	• Serious conditions • Preliminary clinical evidence of substantial gain on clinical end point
Features	• **FDA proactive to expedite development & review, e.g., more FDA assistance meetings & design inputs** • Rolling review option available	• Approval based on surrogate or interim clinical end point • Post-approval trials to verify	• **Shorter clock for approval vs. normal reviews, 6 vs. 10 months** • More FDA accountability	• **All fast track features** • **More FDA trial development guidance** • **Phase 2 data for approvals option** • FDA commitment by senior management

Orphan drugs: [1983] Rare diseases (under 200,000 patients), serious & life-threatening, often children. Benefits – marketing exclusivity for 7 years; FDA research grants; tax credits; application fees waived; trial assistance from FDA. About 40% of NDAs/BLAs (10%–50% per year) are orphan drugs and one-third are biologicals.

Regenerative medicine advanced therapies: [2017] Cell and tissue therapies addressed with accelerated approvals in serious diseases.

Qualified infectious disease product (QIDP): [2018] Accelerated antibiotic product reviews (drugs & biologicals) for serious pathogens. Priority review and marketing exclusivity for 5 years.

Limited population pathway for antibacterial and antifungal drugs: [2018] Rare serious or life-threatening infections in small populations. Patient sample in trials OK to be small.

Abbreviations: BLA – biological license application, FDAMA – Food and Drug Administration Modernization Amendments, FDASIA – Food and Drug Administration Safety and Innovation Act, FDC – Food, Drug, and Cosmetic Act, NDA – new drug application, PDUFA – Prescription Drug User Fee Act.

distribution by a company; (11) *Meetings* of FDA staff with sponsors (companies), usually regarding clinical research or at key milestones in application submission; (12) FDA *Policy announcements* through public symposia or professional meetings; (13) *Postmarketing commitments* for Phase 4 trials and risk evaluation systems examining adverse events; and (14) *Fines*, which can be levied for lack of compliance with regulations.

chapter eight

Biotechnology products and indications I. Proteins

A substantial number of biotechnology products (442) have been developed and marketed by over 120 companies in the United States from the 1980s through 2019 (a 37-year period). In just the last 6 years, about 150 biotechnology products have been developed and marketed. Proteins account for over about 60% of these biotech products. In many diseases, biotechnology products often have been a major breakthrough offering the first effective treatments where previously very little or nothing was effective for serious diseases or now disease progression is moderated; for example, imiglucerase for Gaucher's disease, human papilloma virus vaccine for cervical and vulvar cancer prevention, or beta-interferon for multiple sclerosis, anti-RNA molecules for Duchenne's muscular dystrophy, or gene therapy for spinal muscular atrophy. Furthermore, major advances in the mitigation of chronic immune diseases have been made with biotechnology products impacting the pathogenesis of the disease and not just symptom relief, such as, fusion proteins (e.g., etanercept) and monoclonal antibodies (mabs) (e.g., adalimumab) for arthritis, colitis, and psoriasis. Now, we have biological products in more than 14 distinct categories. Table 8.1 presents a statistical overview of the number of biological molecules, products, indications, and companies as of December 2019. The statistical tally is exceptional. These 442 products for the 367 distinct molecules were created by over 100 originator biotechnology companies, along with their pharma partners, and are used for about 380 separate indications. Following a discussion of each category of biological products in the following narratives, a table of biotechnology products approved for use in the United States presents the generic and brand names, company marketing the product, and therapeutic areas of use.

Table 8.1 Marketed biotech products by categories (as of December 23, 2019)

Types of products	Molecules	Products	Indications[a]	Companies[b]
Proteins/recombinant:				
• Hormones	10	26	31	28
• Enzymes	31	31	23	30
• Growth factors	16	22	19	30
• Interferons	10	13	14	32
• Interleukins	4	4	5	9
• Blood factors	28	30	9	27
• Fusion proteins (Mol. Eng.)	11 (+14)[c]	12 (+15)[c]	20	27
• Toxins (Mol. Eng.)	5	5	15	5
• Monoclonal antibodies	89	109	127	79
Vaccines (Recomb./Mol. Eng.)	27	28	18	18
Peptides (Recomb./Mol. Eng.)	92	105	78	100
Liposomes (Mol. Eng.)	11	13	13	31
Oligonucleotides (RNAi/Gene tx.)	10	10	9	17
Cell therapies	6	6	11	9
Tissue therapies	11	12	5	8
Polymers	9	11	14	22
Others	2	2	2	2
Totals	368	442	380	279 [125][d]

Abbreviations: Mol. Eng. – molecular engineering, Recomb. – recombinant, RNAi – ribonucleic acid inhibition, R&D – research and development, S&M – Sales and marketing, tx – treatment.

[a] The same indication can be listed on multiple product categories.
[b] Companies listed in each category include the original marketing company and any subsequent licensee who acquired the product.
[c] Fusion proteins are listed in other categories as well, as noted by the number in parentheses, e.g., 25 total molecules (11 + 14).
[d] The number of companies (first number) originally involved in R&D and S&M. Number of companies in parentheses that are currently marketing a biotech product.

Proteins – hormones

Protein hormones include 26 products, all created through recombinant DNA (rDNA) technology, as listed in Table 8.2. Indications include primarily deficiencies in hormone production leading to growth conditions, such as, growth failure (children and adults), short stature, AIDS wasting, and acromegaly; hormonal fertility maladies; some insulin-like products

Table 8.2 Protein hormones [n=26] (as of December 21, 2019)

Trade name	Nonproprietary name	Company	Usage area
Growth hormones: [17]			
Accretropin	Somatropin	Emergent Biosol.	Growth deficiency
Bio-Tropin	Somatropin	LifeTech	Growth deficiency
Genotropin	Somatropin	Pfizer	Growth deficiency
Humatrope	Somatropin	Eli Lilly	Growth deficiency
Norditropin	Somatropin	Novo Nordisk	Growth deficiency
Nutropin	Somatropin	Roche	Growth deficiency
Nutropin AQ	Somatropin	Roche	Growth deficiency
Nutropin Depot[a]	*Somatropin*	*Roche/Alkermes*	*Growth deficiency*
Omnitrope	Somatropin	Roche	Growth deficiency
Protropin	*Somatropin*	*Roche*	*Growth deficiency*
Saizen	Somatropin	Merck KGaA	Growth deficiency
Serostim	Somatropin	Merck KGaA	AIDS wasting
Somavert	Pegvisomant	Pfizer	Acromegaly
Tev-Tropin	Somatropin	Ferring	Growth deficiency
Valtropin	*Somatropin*	*LG Life Sciences*	*Growth deficiency*
Zomacton	Somatropin	Ferring	Growth deficiency
Zorbtive	Somatropin	Merck KGaA	Short bowel synd.
Fertility: [3]			
Follistim	Follitropin beta	Merck	Ovulatory failure
Gonal-F	Follitropin alfa	Merck KGaA	Ovulatory failure
Luveris	*Lutropin alfa*	*Merck KGaA*	*Ovulatory failure*
Ovridel	Choriogonadotropin-alfa	Merck KGaA	Ovulatory failure
Other: [3]			
Myalept	Etreleptin	Aegerion	Lipodystrophy
Natpara	Parathyroid hormone	Takeda	Parathyroid defic.
Thyrogen	Thyrotropin	Sanofi	Thyroid cancer
Diabetes mellitus: [3] [combinations, protein + peptide]			
Ozempic	Semaglitide + Albumin	Novo Nordisk	Diabetes mellitus
Tanzeum	*Albiglutide + Albumin*	*GlaxoSmithKline*	*Diabetes mellitus*
Trulicity	Dulaglutide + IgG4Fc	Eli Lilly	Diabetes mellitus

[European Union (EU)/Japan (Jp): Bemfola (Follitropin alfa) – Gedeon Richter, Growject (Somatropin) – Dainippon Sumitomo, Ovaleap (Follitropin alfa) – Teva, Rekovelle (Follitropin delta) – Ferring]

Abbreviations: Defic. – Deficiency, Synd. – Syndrome.

[a] The products in italics have been withdrawn from marketing.

for diabetes mellitus, and hormone deficiencies as in leptin (lipodystrophy), parathyroid, and thyrotropin. Six products have been withdrawn from marketing, identified by italic print in Table 8.2 and in all following product tables. One parent molecule can result in more than one product because manipulations of the molecule can be done while sustaining action of the molecule or patent limitations allowing very similar (almost identical) products to be developed and marketed, such as with the many somatropin products [growth hormone (GH)]. Thirteen GH products are on the market for six indications. Fertility hormones are another common class of biotechnology products, six products in the United States and one additional in Europe. The generic names of products often will be different for similar products produced by different companies, for example, follitropin beta from Merck and follitropin alfa from Merck KGaA-Serono. Osteoporosis therapy involves two recombinant proteins, plus one thyroid cancer treatment, and one product each for acromegaly. Other uses include HIV-related lipodystrophy and other pituitary-type growth deficiencies.

Proteins – enzymes

Protein enzymes include 31 products, as listed in Table 8.3. The first protein enzyme developed in the late 1980s was alteplase (Activase®), a thrombolytic agent used to minimize complications owing to blood coagulation in acute myocardial infarction. Two further enzymes that were molecularly engineered from alteplase were developed for similar indications. The cardiovascular indications related to thrombolysis have expanded to seven, including acute myocardial infarction, pulmonary embolism, stroke, percutaneous angioplasty, arterial vessel stenting, unstable angina, and acute coronary syndromes with or without unstable angina. A unique enzyme was discovered for cystic fibrosis, dornase alfa, which is the enzyme deficiency responsible for the etiology of this disastrous respiratory disease. Cystic fibrotic patients experience prevention of breakdown of DNA from degrading cells in the lungs, causing sludging of alveolar secretions and very poor lung function. This enzyme was the first protein administered by inhalation, given the nature of the disease being a protein deficiency in the alveoli of lungs. A family of rare single-enzyme deficiencies causes life-ending diseases often in childhood or adolescence, involving multiple organ system disruption (blood, liver, and nervous system). Fifteen such enzyme deficiencies have the replacement enzyme commercially available for 12 diseases, that is, Batten's, Fabry's, Gaucher's, Hunter's, Hurler's, Morquio A, Pompe's, Severe compromised immune disease, Sly, and Maroteaux-Lamy, many of which are specific mucopolysaccharidoses, plus the two diseases with hypophosphatemia and lysosomal acid lipase.

Table 8.3 Proteins – enzyme products [n=32] (as of December 31, 2019)

Trade name	Nonproprietary name	Company	Usage area
Cardiovascular: [3]			
Activase	Alteplase	Roche	Thrombolysis, AMI
Retevase	Reteplase	Chiesi	Thrombolysis, AMI
TNKase	Tenecteplase	Roche	Thrombolysis, AMI
Genetic Dis.: [17]			
Adagen	Pegadamase	Leadiant	ADA deficiency
Aldurazyme	Laronidase	Sanofi	Hurler syndrome
Brineura	Cerliponase	BioMarin	Batten disease
Cerezyme	Imiglucerase	Sanofi	Gaucher's dis. type 1
Elaprase	Idursulfase	Takeda	Hunter's disease
Elelyso	Taliglucerase alfa	Pfizer	Gaucher's dis. type 1
Fabrazyme	Algalsidase	Sanofi	Fabry's disease
Kanuma	Sebelipase alfa	Alexion	Lysosomal acid lipase def
Lumizyme	Alglucosidase-2	Sanofi	Pompe's disease
Mepsevii	Vestronidase	Ultragenx	Sly Syndrome
Myozyme	Alglucosidase	Sanofi	Pompe's disease
Naglazyme	Galsulfase	BioMarin	Mucopolysaccharidosis 6
Palynziq	Pegvaliase-pqpz	BioMarin	Phenylketonuria
Revcovi	Elapegademase-lvir	Leadiant	SCID
Strensiq	Asfotase alfa	Alexion	Hypophosphatemia
Vimizim	Elosulfase alfa	BioMarin	Morquio A syndrome
Vpriv	Velaglucerase	Takeda	Gaucher's dis. type 1
Other: [11]			
Asparlas	Calaspargase pegol-mknl	Servier	Lymphoblastic leukemia
Elitek	Rasburicase	Sanofi	Hyperuricemia, cancer
Hylenex	Hyaluronidase	Takeda	Hypodermoclysis
Hyqvia	Hyaluronidase + immunoglobulins	Takeda	Immunodeficiency
Jetrea	Ocriplasmin	Oxurion	Vitreomacular adhesion
Krystexxa	Pegloticase	Horizon	Gout
Oncaspar	peg-l-aspargase	Takeda	Lymphoblastic leukemia
Pulmozyme	Dornase alfa	Roche	Cystic fibrosis
Voraxaze	Glucarpidase	BTG Int'l	Methotrexate toxicity
Xiaflex	Collagenase histolytica	Endo	Duputren's contracture
Xigris[a]	*Drotrecogin A*	*Biocritica*	*Septic shock*
[EU – Replagal (Algalsidase) – AbbVie]			

Abbreviations: ADA – adenosine deaminase, AMI – acute myocardial infarct, Def – deficiency, Dis. – disease, Int'l – international, SCID – severe compromised immune disease.

[a] The products in italics have been withdrawn from marketing.

Novel rDNA technologies were used primarily to create these replacement enzymes, mitigating the diseases; plus, a gene-activated process was used for a recombinant protein (velaglucerase) and a plant host cell for rDNA for taliglucerase, both products are used in Gaucher's disease. Two enzymes are available for uric acid abnormalities. Gout is treated with pegloticase. Also, a key enzyme is responsible for uric acid metabolism, principal to the eradication of excess nucleic acid material from dying cells especially in cancers; its deficiency leads to hyperuricemia, and is correctable with the enzyme product rasburicase. Other quite variable and distinct conditions treated with enzyme therapies are acute lymphoblastic leukemia, primary immunodeficiency, Dupuytren's contracture, methotrexate toxicity, phenylketonuria, and vitreomacular adhesion. Hypodermoclysis (enhancing concurrent drug absorption) involves one enzyme product, Hylenex.

Proteins – growth factors

Growth factor (GF) proteins number 21, as listed in Table 8.4. These proteins can be divided into two areas: blood cell GFs, also known as colony-stimulating factors (CSFs), and tissue GFs, all of which are ligands to communicate between cells and stimulate new cell outcomes and functions. The CSFs are secreted by specific cells in organs, for example, erythropoietin by the kidney, and stimulate another cell type to produce an effect; in this example, bone marrow erythroid progenitors are stimulated to accelerate their production of red blood cells and correct anemia. All these GFs are produced by rDNA technology. CSF products for leukocytes are available worldwide, that is, filgrastim, pegfilgrastim, and sargramostim in the United States, and also molgramostim, regramostim, lenograstim, and nartograstim in rest of the world. Biosimilar filgrastim products include biograstim, filgrastim-Hexal, nivestim, ratiograstim, and tevagrastim. They all stimulate white blood cell production and limit infectious complications in myeloid-suppressed cancer patients. Two epoetin molecules (epoetin alfa and epoetin beta) stimulate red blood cell production, with two U.S. products available (Epogen and Procrit), along with Eprex, NeoRecormon, Silapo and Epogin in Europe and Asia. Biosimilar products for epoetin alfa are also available, such as Retacrit. Aranesp in the United States and Nespo in Europe are the hyperglycosylated products of epoetin that have extended half-lives and require less frequent dosing. A pegylated form of epoetin alfa has been developed as well to extend the dosing interval. Becaplermin is a tissue GF for the epidermis and is used to accelerate wound healing in diabetic ulcers. The second recombinant tissue GF is palifermin, impacting keratinocytes, and is used to more rapidly resolve the mucositis in cancer patients receiving chemotherapy, radiation therapy, and bone marrow

Table 8.4 Proteins – growth factors [n=21] (as of Decmber 31, 2019)

Trade name	Nonproprietary name	Company	Usage area
Aranesp	Darbepoetin	Amgen	Anemias
Augment bone graft	PDGF-BB	Wright Med. Grp.	Fusion, ankle
Epogen	Epoetin alfa	Amgen	Anemias
Fulphila	Peg-filgrastim-jmdb	Mylan/Biocon	Neutropenia cancer
GEM 21S	PDGF-BB	Wright Med. Grp.	Periodontal
Infuse	BMP-2	Pfizer/Medtronic	Bone fusion
Kepivance	Palifermin	SOBI	Mucositis, cancer
Leukine	Sargramostim	Partner Therap.	Neutropenia, cancr
Mircera	Epoetin-beta-peg	Vifor Pharma	Anemias
Neulasta	Filgrastim-peg	Amgen	Neutropenia, cancr
Neupogen	Fligrastim	Amgen	Neutropenia, cancr
Neutroval	Filgrastim tbo	Teva	Neutropenia, cancr
Nivestym	Filgrastin-aafi	Pfizer	Neutropenia, cancr
OP-1 Protein	Osteogenic protein-1	Mariel Therap.	Bone reunion
Oxervate	Cenegermin	Dompe	Neurotro. keratitis
Procrit	Epoetin alfa	Johnson & Jhnsn	Anemias
Regranex	Becaplermin	Smith-Nephew	Foot ulcer
Retacrit	Epoetin-alfa-epbx	Pfizer	Anemias
Udencya	Pegfilgrastim-cbqv	Coherus BioSci.	Neutropenia, cancr
Zarxio	Filgrastim-sndz	Novartis	Neutropenia, cancr
Ziextenzo	Pegfilgrastim-bmez	Novartis	Neutropenia, cancr

[EU-Japan: G-Lasta (Filgratim-peg) – Kyowa-Kirin, Gran/Neu-up (Filgrastim) – Kowa-Hako-Kirin, Granix (Filgrastim) – Teva, NeoRecormon/Epogin (Epoetin alfa) – Chugai, Neutrogin/Granocyte (Filgrastim) – Chugai, Silapo (Epoetin zeta) – Stada Arzneimittel]

Abbreviations: BioSci – Biosciences, Cancr – Cancer, Jhnsn – Johnson, Med. Grp. – medical group, Neurotro. – neurotrophic, PDGF – platelet derived growth factor, Therap. – therapeutics.

transplants. Bone growth and bone fusion are accelerated with osteogenic protein-1 and platelet derived GF-BB, both GFs. Recently, neurotrophic keratitis has become treatable with a GF, oxervate. Pegylation has been used to create new GFs with longer half-lives and extended duration of action, for example, filgrastim daily is dosed daily for 5–10 d versus peg-filgrastim (Neulasta) in a single dose. Biosimilar products have also been approved for filgratim around the world; Fulphila, Nivestym, Udencya, and Zarxio.

Proteins – interferons and interleukins (cytokines)

The cytokine proteins number 13 interferons and 4 interleukins, as listed in Table 8.5. The proteins called interferons are produced by many cells in the human body and participate in our immune system to provide protection against foreign substances, such as infectious material or cancer, and are involved in immune diseases. Interferons have several beneficial properties; an indirect mechanism of action, which stimulates the immune system (more immunity), especially lymphocytes; and also possess direct cytolytic and antiviral activity. Three families of interferons exist (alpha, gamma, and beta) and 13 products have been developed (8-alfa, 1-gamma, and 4-beta). The indications (11) for interferons are broad, including oncology (e.g., malignant melanoma, hairy cell leukemia, Kaposi's sarcoma, follicular lymphoma, and chronic myelogenous leukemia), viral infection (e.g., hepatitis B and C, genital warts), and autoimmune neurologic disease (e.g., multiple sclerosis). Beta interferons at the time revolutionized the treatment of multiple sclerosis, slowing neurologic disease progression as well as reducing the debilitating effects of the disease. Newer molecular forms include pegylated molecules

Table 8.5 Proteins – cytokines – interferons and interleukins [n=17] (as of December 31, 2019)

Trade name	Nonproprietary name	Company	Usage area
Actimmune	Interferon-gamma-1b	Clinigen	Granulomatous dis.
Avonex	Interferon-beta-1a	Biogen	Multiple sclerosis
Betaseron	Interferon-beta-1b	Bayer	Multiple sclerosis
Extavia	Interferon-beta-1b	Novartis	Multiple sclerosis
Infergen	Interferon-alfacon-1	Kadmon	Hepatitis C
Intron-A	Interferon-alfa-2b	Merck	Hairy cell leukemia
Kineret	Anakinra, IL-1a	SOBI	Hepatitis C
Neumega	Oprelvekin	Pfizer	Thrombocytopenia
Ontak	Denileukin diftitox	Eisai	T-cell lymphoma
Pegasys	Interferon-alfa-2a peg	Roche	Hepatitis C & B
Peg-Intron	Interferon-alfa-2b peg	Merck	Hepatitis C
Plegridy	Interferon-beta-1a peg	Biogen	Multiple sclerosis
Proleukin	Aldesleukin	Clinigen	Kidney & skin cancer
Rebetron	Interferon-alfa-2b+Ribivirin	Merck	Hepatitis C
Rebif	Interferon-beta-1a	Merck KGaA	Multiple sclerosis
Roferon-A[a]	*Interferon-alfa-2a*	*Roche*	*Hairy cell leukemia*
Sylatron	Interferon-alfa-2b	Merck	Melanoma

Abbreviation: Dis – disease.

[a] The products in italics have been withdrawn from marketing.

that provide the advantages of longer duration of action and thus require less frequent dosing while sustaining the antiviral activity. Indications often are added to a product over time following the original product approval. Alpha-interferon is a very good example of this drug development process. The first indication was hairy cell leukemia in 1986, a narrow use for an uncommon severe oncologic problem, which was followed by approval for 10 additional uses over the subsequent 20 years. Extensive clinical trials (Phases 2 and 3) and a supplemental new drug application were required to establish the safety, efficacy, and appropriate dosing for each use. In Europe, additional interferon products are marketed and even more in Asia. Hepatitis is a much more common disease in Asia and Europe.

Four interleukins are in use for renal cell carcinoma and malignant melanoma [interleukin-2 (IL-2)], cutaneous T-cell lymphoma (denileukin), thrombocytopenia associated with cancer chemotherapy (IL-11), and rheumatoid arthritis (IL-1ra, anakinra). These interleukins are protein products from rDNA manufacturing and physiologically are usually local intercellular communication molecules that, when used systemically as a therapeutic, can cause substantial multi-organ toxicity, especially cardiovascular, limiting their full clinical usefulness and characterizing most interleukins. Denileukin is a fusion protein of IL-2 and diphtheria toxin. Anakinra is a receptor antagonist to IL-1. Oprelvekin (IL-11) stimulates progenitor cells such as megakaryocytes to produce more platelets and reduce bleeding in several conditions.

Proteins – blood coagulation factors

The coagulation proteins number 30, as listed in Table 8.6. Six companies market most of the products worldwide, that is, Bayer, CSL Behring, Novo Nordisk, Octapharma, Pfizer, and Takeda. Blood factors are very large proteins (thousands of amino acids in two chains [- A & B]) and are involved in normal blood coagulation as cofactors in the coagulation cascade [Factor 7, F.8, F.9, F.10, F.13, and thrombin, plus von Willebrand factor]. The deficiency of any one blood factor, often a genetic deficiency, leads to serious bleeding disorders (hemophilia), but it is fully correctable through replacement therapy with these rDNA proteins. The factors are recombinant DNA proteins employing two possible hamster host cells, baby kidney (BKH) or Chinese ovaries (CHO). Factor 8 is available in 18 products, all with the same indication and use, and Factor 9 in 6 products. The original blood factors were short acting, but the use of pegylation (Adynovate, Esperoct, Jivi, Rebinym) and fusion proteins (e.g., Afystyla, Alprolix) extends the products' half-lives substantially. Also, further molecular engineering of the protein domains, mostly B-domain deletion or truncation, creates additional molecules and products; however, the indications are the same as for the originator product. Other coagulation proteins are marketed for Factor 7 and von

Table 8.6 Proteins – coagulation factors [n=8.6] (as of December 31, 2019)

Trade name	Nonproprietary name	Company	Usage area
Advate	Octocog-alfa. F.8	Takeda	Hemophilia A
Adynovate	Damoctocog-alfa, F.8 peg	Takeda	Hemophilia A
Afstyla	Factor 8	CSL Behring	Hemophilia A
Alprolix	F.9+IgG1Fc	Bioverative	Hemophilia B
Andexxa	Factor Xa inactive-zhzo	Portola	Factor X antidote
Atryn	Antithrombin, goats	Kadmon	Antithrombin defic.
BeneFix	Factor 9, Nonacog-alfa	Pfizer	Hemophilia B
Bioclate	Factor 8	CSL Behring	Hemophilia A
Eloctate	F.8 A-domain+IgG1Fc	Bioverative	Hemophilia A
Esperoct	Turoctocog-pegol	Novo Nordisk	Hemophilia A
Helixate FS	Octocog-alfa, F.8	CSL Behring	Hemophilia A
Idelvion	Albutrepanoctocog-alfa, F.9	CSL Behring	Hemophilia B
Ixinity	Nonacog-alfa, F.9	Aptevo Therap.	Hemophilia B
Jivi	Damoctacog alfa pegol	Bayer	Hemophilia A
Kogenate	Factor 8	Bayer	Hemophilia A
Kogenate FS	Factor 8	Bayer	Hemophilia A
Kovaltry	Factor 8	Bayer	Hemophilia A
NovoEight	Turoctocog-alfa, F.8	Novo Nordisk	Hemophilia A
NovoSeven	F.7, Eptacog-alfa	Novo Nordisk	Hemophilia F.7
Nplate	Romiplostim	Amgen	Thrombocyt. purpura
Nuwig	Simoctocog-alfa, F.8	Octapharma	Hemophilia A
Obizur	Octocog-alfa, F.8	Takeda	Hemophilia A
Rebinyn	Nonacog-beta-peg, F.9	Novo Nordisk	Hemophilia B
Recombinate	Factor 8	Takeda	Hemophilia A
Recothrom	Thrombin	Mallinckrodt	Bleeding, surgery
ReFacto	Moroctocog-alfa, F.8	Pfizer	Hemophilia A
Rixubis	Nonacog-alfa, F.9	Takeda	Hemophilia B
Tretten	Factor 13	Novo Nordisk	Hemophilia F.13
Vonvendi	von Willebrand factor	Takeda	von Willebrand dis
Xyntha	Octocog-alfa, F.8	Pfizer	Hemophilia A

[E.U. – Adynovi (Rurioctocog-alfa, F.8 peg) – Takeda, Vihuma (Simoctocog-alfa) – Octapharma]

Abbreviations: Defic – deficiency, Dis – disease, F – factor, Peg – pegylation, Therap. – therapeutics, Thrombocyt. – thrombocytopenic.

Willebrand deficiencies. These recombinant protein blood factors replace natural blood derivative products and avoid the potential viral contamination and immune reactions that were previously observed in these patients receiving blood derived products. Additionally, both a thrombin and anti-thrombin protein are marketed. An antidote to blood thinners is available as well, Andexxa. One coagulation protein (Atryn) is an antithrombin that involves manufacture in transgenic goats to produce the protein in their milk. Thrombocytopenia purpura has a protein therapeutic available (Nplate).

Proteins – toxins

Botulinum toxins from the microbe *Clostridium botulinum* are employed in a wide range of conditions, primarily utilizing its muscle-paralyzing activity; they alter muscle function in various situations such as cervical dystonia, upper and lower limb spasticity, spasticity in cerebral palsy-stroke-trauma, strabismus, and blepharospasm; also, sialorrhea and axillary hydrosis; plus urinary incontinence in neurologic diseases; and, additionally, improvement in glabellar lines, forehead lines, and crow's feet in cosmetic surgery; and migraine headaches and multiple sclerosis. Five products have been approved for use from seven companies, as noted in Table 8.7.

Fusion protein products

In the prior discussion of molecular engineering, two separate biologically active entities are merged structurally into one protein-related product to achieve a new set of combined properties. Fusion molecules are comprised most often of a combination of two molecules, such as, antibody fragments, protein receptors, conjugate molecules (toxins, drugs, or radionuclides), and peptides. Proteins most commonly are employed for at least one of the two components. Thirty fusion molecules are now marketed, as listed in Table 8.8. We will discuss several vaccines that also are fusion molecules and which

Table 8.7 Proteins – botulinum toxins [n=5] (as of December 31, 2019)

Trade name	Nonproprietary name	Company	Usage area
Botox	Onabotulinum toxin A	Allergan	Various
Dysport	Anobotulinum toxin A	Ipsen/Galderma	Various
Jeuveau	Parbotuminum toxin A-xyfs	Evolus	Glabellar lines
Myobloc	Rimaboltulinum toxin B	Alkermes/Solstice	Cervical dystonia
Xeomin	Inocobotulinum toxin A	Merz	Various

Notes: Various indications – cosmetic in facial lines, dystonias (limbs-lower & upper, eyelids, cervical), hyperdrosis/axillary, migraine, multiple sclerosis, sialorrhea, urinary incontinence.

Table 8.8 Fusion proteins [v=13 + 17 other categories + 2 EU & 1 Jp]
(December 31, 2019)

Trade name	Nonproprietary name	Company	Usage area
Immune disorders: [8]			
Abraxane	Paclitaxel-albumin	BristolMyrsSqbb	Cancer, 4 organs
Amevive[a]	*Alefacept*	*Biogen-Idec*	*Psoriasis*
Enbrel	Etanercept	Amgen/Pfizer	Ps/UC, Cr/RA, AS
Erelzi	Etanercept-szzs	Novartis	Ps/UC, Cr/RA, AS
Eticovo	Etanercept-ykro	Samsung Bioepis	Ps/UC, Cr/RA, AS
Eylea	Aflibercept	Baxter	Macular degen.
Nulojix	Belatacept	BristolMyrsSqbb	Kidney rejection
Orencia	Abatacept	BristolMyrsSqbb	Arthritis
Other: [5]			
Adagen	Pegadamase	Leadiant Biosci.	SCID
Arcalyst	Rilonacept	Regeneron	CAPS
Elzonris	Tagaxofusp-erzs	Stemline Therap.	Dendritic cell can
Reblozyl	Lustatercept-aamt	Celgene	Beta-thalassemia
Zaltrap	Aflibercept	Sanofi	Colorectal cancer

Fusion proteins in other categories:

Trade name	Proprietary name or peptides/proteins	Marketing company
Diabetes mellitus: [3]		
Ozempic	Semaglutide + Albumin	Novo Nordisk
Rybelsus	Semaglutide + Albumin + Fatty diacid	Novo Nordisk
Tanzeum	*Albiglutide + Albumin*	*GlaxoSmithKline*
Trulicity	Dulaglutide + IgG4Fc	Eli Lilly
Blood disorders: [3]		
Alprolix	Factor 9 + IgG1Fc	Biogen/SOBI
Eloctate	Efraloctocog-alfa, Factor 8 A-domain + IgG1Fc	Biogen/SOBI
Nplate	Romiplostim, 2 Fc fragments + 2 thrombopoietin receptor peptides	Amgen
Cytokines – cancer: [1]		
Ontak	Denileukin diftitox	Eisai
Enzymes: [1]		
Strensiq	Asfotase alfa	Alexion
Vaccines: [6]		
Bexsero	2 Meningococcal B proteins + Factor H binding protein	GlaxoSmithKline

(Continued)

Table 8.8 **(Continued)** Fusion proteins [n=13+17 other categories + 2 EU & 1 Jp] (December 31, 2019)

Trade name	Proprietary name or peptides/proteins	Marketing company
Menactra	4 Meningococcal polysaccharide + Diphtheria toxin	GlaxoSmithKline
Menveo	4 Meningococcal polysaccharide + Diphtheria CRM_{197}	GlaxoSmithKline
Prevnar-7	7 Pneumococcal serotypes + Diphtheria CRM_{197} protein	Pfizer
Prevnar-13	13 Pneumococcal serotypes + Diphtheria CRM_{197} protein	Pfizer
Trumemba	2 Meningococcal B serotypes + Factor H binding protein	Pfizer

Mab conjugate: [2]

Adcetris	Brentuximab vedotin	Seattle Genetics
Lumoxiti	Moxetumomab pasotox-tdfk	AstraZeneca

[E.U./Jp – Benepali (Etanercept) – Biogen, Lifmior (Etanercept) – Pfizer, Romiplate – Kyowa-Hakko-Kirin]

Pegylated Products: [34]

Peg-GFs: [5] Peg-fligrastim (Neulasta; Fulphila; Udenyca); Peg-epoetin beta (Mircera), *Peginesatide (Omontys)*

Peg-Enzymes: [6] Pegademase (Adagen), Peg-asparagase (Oncaspar), Pegloticase (Krystexxa), Pegvaliase-pqpz (Palynziq), Elapegademase-lvir (Revcovi), Calasparagase pegol-mknl (Asparlas)

Peg-Hormone: [1] Pegvisomant (Somavert)

Peg-Interferons: [4] Peg-Interferon-2a (Pegasys), Peg-Interferon-2b (Peg-Intron; Sylatron); Peg-Interferon-beta-1a (Plegridy)

Peg-Coagulation proteins: [2] Peg-Factor 8 (Adynovate), Peg-Factor 9 (Reninyn)

Peg-mab: [1] Certolizumab (Cimzia)

Peg-Aptamer: [1] Pegaptamib (Macugen)

Peg-Liposomes: [14] Amikacin-liposomal (Arikayce), Amphotericin-liposomal (Abelcet, Amphotec, AmBisome), Bupivacaine-liposomal (Exparel), Cytarabine-liposomal (DepoCyt), Daunomycin + Cytarabine (Vyxeos), Doxorubicin-liposomal (DaunoXome, Doxil, Caelyx, LipoDox), Irinotecan-liposomal (Onivyde), Morphine-liposomal (Depodur), Vincristine-liposomal (Marqibo)

Abbreviations: AS – ankylosing spondylitis, BioSci – BioSciences, BristolMyrsSqbb – Bristol-Myers-Squibb, Can – cancer, CAPS – cryoprin-associated periodic fever, Cr – Crohn's disease, degen – degenerative, Ps – psoriasis, RA – rheumatoid arthritis, SCID – severe compromised immune disease, Therap. – therapeutics, UC – ulcerative colitis.

[a] The products in italics have been withdrawn from marketing.

incorporate an antigenic microbial protein or polysaccharide and an immunity booster peptide or protein toxic compound, for example, Menactra (diphtheria toxin used for meningococcal infection) and Prevnar (polysaccharide used for pneumococcal infection). Several antibody fragments, usually the Fc structural fragment, are combined with antigens or proteins representative of a disease pathology (most often immune disorders), permitting mitigation of the disease. For example, etanercept (Fc antibody fragment and the extracellular domain of tumor necrosis factor receptor, a derivative of the primary inflammatory protein) is used for rheumatoid arthritis, colitis, and psoriasis. Abatacept for rheumatoid arthritis, and rilonacept for periodic cryopyrin-associated periodic disease also use the Fc mab fragment. As an example of the disruption of disease pathogenesis, the toxic tumor necrosis factor causing inflammation and tissue destruction is blocked by etanercept to improve rheumatoid arthritis pathology and signs of the disease. The indications of fusion proteins were extended to cover several inflammatory conditions involving varied organs, such as ankylosing spondylitis, Crohn's disease, cryoprin-associated periodic disease, psoriasis, psoriatic arthritis, and ulcerative colitis, plus acute macular degeneration, beta-thalassemia, colorectal cancer, and kidney transplant rejection.

An interleukin derivative in oncology, an albumin product used in oncology, a pegylated aptamer, and a platelet-active product are fusion molecules as well. In diabetes mellitus, peptides are combined with proteins to extend the duration of the product's action. Blood disorders are also treated with fusion proteins achieving again longer product actions. Two mabs are antibody-drug-conjugates used in oncology (Adcetris and Lumoxiti). In a more broad description of fusion molecules, the pegylated products are a fusion of polyethylene glycol (PEG) molecules with a protein or peptide or liposome, and there are 34 such products (4 interferons, 5 GFs, 6 enzymes, 2 coagulation factors, 1 hormones, 1 mab, 1 aptamer, 14 liposomes). Pegylation of the liposomal molecules offers protection versus the immune system, adding a safety feature.

Monoclonal antibodies

The 1990s were called initially the biological era of mab proteins, as one product was approved in 1980s that expanded to seven products in the 1990s with a wide range of indications. However, this growth continued much more dramatically forward into the twenty-first century with 20 more products in the 2000s and 80 in the 2010s (up to through December 2019). Table 8.9 lists the 108 mabs with their marketing company (38 companies) and usage areas (90 uses); some mabs are listed in more than one disease category related to their multi-organ multi-disease activity. The nomenclature for mabs is highly structured and guides the identification of their origin and usage areas. For example,

Table 8.9 Proteins – monoclonal antibodies [n=109] (as of December 31, 2019)

Trade name	Nonproprietary name	Company	Usage area
Cardiology: [5 products – 4 uses]			
Actemra	Toclizumab	Roche	Giant cell arteritis
Praluent	Alirocumab	Sanofi & Regeneron	Hypercholesterol.
Praxbind	Idarucizumab	Boehringer-Ingel.	Drug Antidote
ReoPro	Abciximab	Eli Lilly	Thrombolytic
Repatha	Evolocumab	Amgen	Hypercholesterol.
Dermatology: [22 products – 5 uses]			
Abrilada	Adalimumab-afzb	Pfizer	Psoriasis
Amjevita	Adalimumab-atto	Amgen	Psoriasis
Avisola	Infliximab-axxq	Amegn	Psoriasis
Cosentyx	Secokinumab	Novartis	Psoriasis
Cyltezo	Adalimumab-adbm	Boehringer-Ingel.	Psoriasis
Dupixent	Dupilumab	Sanofi & Regeneron	Atop. derm.
Hadlima	Adalimumab-bwwd	Samsung Bioepis	Psoriasis
Humira	Adalimumab	AbbVie	Psoriasis
Hyrimoz	Adalimumab-adaz	Roche	Psoriasis
Ilumya	Tildralizumab	Sun Pharma	Psoriasis
Inflectra	Infliximab-dyyb	Pfizer & Celltrion	Psoriasis
Ixifi	Infliximab-qbtx	Pfizer	Psoriasis
Raptiva[a]	*Efalizumab*	*Roche*	*Psoriasis*
Remicade	Infliximab	Merck & Johnson & Johnson	Psoriasis
Renflexis	Infliximab-abda	Merck & Samsung Bioepis	Psoriasis
Siliq	Brodalumab	Bausch Health	Psoriasis
Skyrizi	Risankizumab-rzaa	AbbVie	Psoriasis
Stelara	Ustekinumab	Johnson & Jhnsn	Psoriasis
Takhzyro	Lanadelumab-flyo	Takeda	Angioedema
Taltz	Ixikizumab	Eli Lilly	Psoriasis
Tremfya	Guselkumab	Johnson & Jhnsn	Psoriasis
Xolair	Omalizumab	Roche & Novartis	Urticaria
Gastroenterology: [16 products – 2 uses]			
Abrilada	Adalimummab-afzb	Pfizer	Colitis, Crohn's
Amjevita	Adalimumab-atto	Amgen & Allergn	Colitis, Crohn's
Avsola	Infliximab-axxq	Amgen	Colitis, Crohn's
Cimzia	Certolizumab pegol	UCB	Crohn's
Cyltezo	Adalimumab-adbm	Boehringer-Ingel.	Colitis, Crohn's

(*Continued*)

*Table 8.9 (**Continued**)* Proteins – monoclonal antibodies [n=109]
(as of December 31, 2019)

Trade name	Nonproprietary name	Company	Usage area
Entyvio	Vedolizumab	Takeda	Colitis, Crohn's
Hadlima	Adalimumab-bwwd	Samsung Bioepis	Colitis, Crohn's
Humira	Adalimumab	Abbvie	Colitis, Crohn's
Hyrimoz	Adalimumab-adaz	Roche	Colitis, Crohn's
Inflectra	Infliximab-dyyb	Pfizer & Celltrion	Colitis, Crohn's
Ixifi	Infliximab-qbtx	Pfizer	Colitis, Crohn's
Remicade	Infliximab	Merck & Johnson & Johnson	Colitis, Crohn's
Renflexis	Infliximab-abda	Merck & Samsung Bioepis	Colitis, Crohn's
Simponi	Golimumab	Merck & Johnson & Johnson	Colitis
Stelara	Ustekinumab	Johnson & Jhnsn	Crohn's
Tysabri	Natalizumab	Biogen	Crohn's
Gynecology: [2 products – 1 use]			
Evenity	Romosozumab-aqqg	Amgen & UCB	Osteoporosis
Prolia	Denosumab	Amgen	Osteoporosis
Hematology: [6 products – 10 uses]			
Adakveo	Crizanlizumab-tmca	Novartis	VOC, Sickle cell
Cablivi	Caplacizumab-yhdp	Sanofi	Thrombocyt. purp.
Gamifant	Emapalumab-lzsg	SOBI	Lymphohistiocyt.
Hemlibra	Emicizumab	Roche	Hemophilia A
Ilaris	Canakinumab	Novartis	CAPS, FMF, #5 dis
Ultomiris	Ravulizumab-cwvz	Alexion	PNH
Infectious disease: [5 products – 4 uses]			
Abthrax	Raxibacumab	Emergent Biosol.	Anthrax
Anthim	Obiltoxaximab	Elusys	Anthrax
Synagis	Pavivizumab	Astrazeneca & SOBI	RSV
Trogarzo	Ibilizumab	TheraTechnologies	HIV
Zinplava	Bezlotoxumab	Merck	*Clostridium difficile*
Neurology: [8 products – 4 uses]			
Aimovig	Erenumab-aooe	Amgen & Novartis	Migraine headache
Ajovy	Fremanuzumab-vfrm	Teva	Migraine headache

(*Continued*)

Table 8.9 (Continued) Proteins – monoclonal antibodies [n=109]
(as of December 31, 2019)

Trade name	Nonproprietary name	Company	Usage area
Emgality	Galcanezumab-gnlm	Eli Lilly	Migraine headache
Lemtrada	Alemtuzumab	Sanofi	Multiple sclerosis
Ocrevus	Ocrelizumab	Roche & Biogen	Multiple sclerosis
Soliris	Eculizumab	Alexion	PNH, HUS
Tysabri	Natalizumab	Biogen	Multiple sclerosis
Zynbryta	*Daclizumab*	*AbbVie & Biogen*	*Multiple sclerosis*
Oncology: [48 products – over 40 cancers]			
Adcetris	Brentuximab vendotin	Takeda	Lymphomas
Arzerra	Ofatumumab	Novartis	Lympho. leukemia
Avastin	Bevacizumab	Roche	OV, CRC, 8 organs
Bavencio	Brodlumab	Merck KGaA & Pfizer	RCC, Blddr, Mrkl
Bespona	Ontotuzumab ozogamicin	Pfizer	Lympho. leukemia
Bexxar	*Tositumomab I-121*	*GlaxoSmithKline*	*Follicular NHL*
Blincyto	Blinatumomab	Amgen	B-cell lymph. Leuk
Campath	*Alemtuzumab*	*Sanofi*	*B-cell Lymph. Leuk*
Cyramza	Ramicizumab	Eli Lilly	GE, NSCLC, CRC
Darzalex	Daratumumab	Johnson & Jhnsn	Multiple myeloma
Empliciti	Elotuzumab	AbbVie & Bristol MyersSquibb	Multiple myeloma
Erbitux	Cetuximab	BristolMyrsSqubb	H/N, CRC
Gazyva	Obinutuzumab	Roche & Biogen	Lymphomas
Enhertu	Fam-trastuzumab deruxtecan-nxki	D-S/AZ	Breast cancer
Herceptin	Trastuzumab	Roche	Breast, GE
Herceptin-hyclea	Traztuzumab hyaluronidase	Roche	Breast, GE
Herzuma	Trastuzumab-pkrb	Teva & Celltrion	Breast, GE
Imfinzi	Durvalumab	AstraZeneca	Uro, NSCLC
Kadcyla	Ado-trastuzumab etamsine	Roche	Breast
Kanjinti	Tratuzumab-anns	Amgen & Allrgn	Breast, GE
Keytruda	Pembrolizumab	Merck	13 organs

(Continued)

Table 8.9 (Continued) Proteins – monoclonal antibodies [n=109]
(as of December 31, 2019)

Trade name	Nonproprietary name	Company	Usage area
Lartruvo	Olaratumab	Eli Lilly	Sarcoma
Libtayo	Cemiplimab-rwlc	Sanofi & Regnrn.	Cutaneous
Lumoxiti	Moxetumomab pasudotox-tdfk	AstraZen.	Hairy cell leuk.
Mvasi	bevacizumab-awwh	Amgen & Allrgn.	OV, CRC, 8 organs
Mylotarg	Gemtuzumab ozogamicin	Pfizer	Myeloid leuk CD33
Ogivri	Trastuzumab-dkst	Mylan & Biocon	Breast, GE
Ontruzant	Trastuzuma-dttb	Merck & Samsung Bioepis	Breast, GE
Opdivo	Nivolumab	BrstlMyrsSqubb	8 organs
Padcev	Enfortumab vedotin-ejfv	Seat. Gen./Ast.	Urothel. Ca
Perjeta	Pertuzumab	Roche	Breast
Polivy	Polatuzumab vedotin	Roche/Seattle Gen.	DLBCL
Poteligeo	Mogamulizumab-kpkc	Kyowa-Hkk-.Krn.	Mycoses fungoides
Portrezza	Necitumumab	Eli Lilly	NSCLC
Rituxan	Rituximab	Roche & Biogen	Lymphomas +6
Rituxan-hycela	Rituximab-dttb	Merck/Samsung Bioepis	Lymphomas
Ruxience	Rituximab-pvvr	Pfizer	Lymphomas +6
Sylvant	Siltuximab	Jazz	Castleman's dis.
Tecentriq	Atezolizumab	Roche	Uro +3 organs
Trazimera	Trastuzumab-qyyp	Pfizer	Breast, GE
Truxima	Rituximab-abbs	Teva & Celltrion	Lymphomas +6
Unituxin	Denutuxumab	United Therap.	Neuroblastoma
Vectibix	Panitumumab	Amgen	CRC
Xgeva	Denosumab	Amgen	Bone rxns in cancr
Yervoy	Ipilimumab	BrstlMyrsSqbb	Skin, CRC, RCC
Zevalin	Ibritumomab In-111	Acrotech	B-cell NHL
Zirabev	Bevacizumab-dbll	Pfizer	Ovarian cancer (OC), CRC ++

Ophthalmology: [2 product – 6 uses]

Lucentis	Ranibizumab	Roche & Novartis	Macular dgnrtn #6
Beovu	Brolucizumab-dbll	Novartis	Macular dgnrtn

(Continued)

Table 8.9 (Continued) Proteins – monoclonal antibodies [n=109]
(as of December 31, 2019)

Trade name	Nonproprietary name	Company	Usage area
Pulmonary: [4 – 2 uses]			
Cinqair	Rezlizumab	Teva	Asthma eosinoph.
Fansenra	Benralizumab	AstraZeneca	Asthma esosinoph.
Nucala	Mepolizumab	GlaxoSmithKline	Asthma eosinoph.
Xolair	Omalizumab	Roche& Novartis	Asthma
Rheumatology: [19 products – 9 uses]			
Abrilada	Adalimumab-afza	Pfizer	AS, JRA, PsA, RA
Actemra	Toclizumab	Roche	Arthritis
Amjevita	Adalimumab-atto	Amgen	AS, JRA, PsA, RA
Avsola	Infliximab-axxq	Amgen	RA, PsA
Benlysta	Belimumab	GlaxoSmithKline	SLE
Cimzia	Certolizumab pegol	UCB	AS, PsA, RA
Cosentyx	Secokinumab	Novartis	AS, PsA
Cyltezo	Adalimumab-adbm	Boehringer-Ingel.	AS, JRA, PA, RA
Hadlima	Adalimumab-bwwd	Samsung Bioepis	AS, JRA, PA, RA
Humira	Adalimumab	AbbVie	AS, JRA, PsA, RA
Hyrimoz	Adalimumab-adaz	Roche	AS, JRA, PsA, RA
Ilaris	Canakinumab	Novartis	CAPS, FMF, 5 dis.
Inflectra	Infliximab-dyyb	Pfizer & Celltrion	RA, PsA
Ixifi	Infliximab-qbtx	Pfizer	RA, PsA
Kevzara	Sarilumab	Sanofi & Regnrn	Rheumat. arthritis
Remicade	Infliximab	Merck & Johnson & Johnson	RA, PsA
Renflexis	Infliximab-abda	Merck & Samsung Bioepis	RA, PsA
Rituxan	Rituximab	Roche & Biogen	Rheumat. arthritis
Simponi	Golimumab	Merck & Johnson & Johnson	AS, PsA, RA
Transplant: [3 products – 3 uses]			
Orthoclone OKT3	*Muromomab-CD3*	*Johnson & Jhnsn*	*Kidney, Liver, Heart*
Simulect	Basiliximab	Novartis	Kidney, Liver, Heart
Zenapax	*Daclizumab*	*Roche*	*Kidney*

(Continued)

Table 8.9 (*Continued*) Proteins – monoclonal antibodies [n=109]
(as of December 31, 2019)

Trade name	Nonproprietary name	Company	Usage area
Other: [1 product – 1 use]			
Crysvita	Burosumab	Ultragenx & KHK	Hypophosphatem.
[E.U. Imraldi (Adalimumab)-Samsung Bioepis, Rixathon (Rituximab-dkst) – Novartis]			

Note: A monoclonal antibody can be listed in multiple medical disciplines related to its broad indications.

Company names: Allrgn – Allergan, Ast – Astellas, AZ – AstraZeneca, Brstl-Myrs-Squibb – Bristol-Myers-Squibb, D-S – Daiichi-Sankyo, Gen. – Genetics, Ingel – Ingelheim, Jhnsn–Johnson, KHK–Kyowa-Hakko-Kirin, Regnrn–Regeneron, Seat. Gen. – Seattle Genetics, Therap. – Therapeutics.

Diseases: AS – ankylosing spondylitis, Atop. Derm. – atopic dermatitis, Ca or can or cancr – cancer, Blddr – bladder cancer, CAPS – cryoprin-associated periodic fever, CRC – colorectal cancer, Dgnrtn – degeneration, Dis – disease, DLBCL – diffuse large B-cell lymphoma, Eosinoph – eosinophilic, FMF – familial Mediterranean fever, GE – gastroesophageal, HIV – human immunodeficiency virus, H/N – head and neck cancer, HUS – hemolytic uremic syndrome, Hypercholest. – hypercholesterolemia, JRA – juvenile rheumatoid arthritis, Leuk – leukemia, Lympho – lymphocytic, Mrkl – Merkle cell carcinoma, NHL – non-Hodgkin's lymphoma, NSCLC – non-small cell lung cancer, OV – ovarian cancer, PNH – paroxysmal nocturnal hemoglobinuria, PsA – psoriatic arthritis, RA – rheumatoid arthritis, Rheumat. – rheumatoid, RCC – renal cell carcinoma, RSV – respiratory syncytial virus, Rxns – reactions, SLE – systemic lupus erythematosus, Thrombocyt. pur. – thrombocytopenic purpura, Uro or urothel– urothelial cancer, VOC – veno-occlusive crisis in sickle cell disease.

[a] The products in italics have been withdrawn from marketing.

trastuzumab (Herceptin) can be identified as used in cancer and is a "humanized" mab as follows: tras-tu-zu-mab; "tras" is the name fragment unique to this mab protein; "tu" is identified for cancer indications, "zu" identifies the "humanized" type (90%) of mab protein, and "mab" of course is for monoclonal antibody. Other common abbreviations include "li" for inflammatory conditions, "lu" for oncology also, and "ci" for a cardiovascular indications or mechanisms of action, "mo" for fully murine proteins, "xi" for chimeric proteins (75% human and 25% murine), and "mu" for fully human (100%) mabs.

Mabs are very highly selective and specific in their targets, attaching to a single cell surface receptor (protein) that usually manifests in a disease. Cancer and immune diseases have many identifiable cell receptors individualized for the diseases and target organs and hence have become the two largest disease categories for mab utilization. In addition to treatment of rejection of organ transplants, mab indications are quite wide-ranging and include prevention of blood clots, 40 cancer types (plus numerous subtypes), for example, metastatic breast cancer, multiple myeloma, several non-Hodgkin's lymphomas, leukemias, colorectal, head, and neck cancers; 17 inflammatory diseases especially of the gastrointestinal, rheumatology, and dermatology areas, for example, ankylosing spondylitis, Crohn's disease, rheumatoid arthritis, psoriasis, and ulcerative colitis; neurologic disease, for example, multiple sclerosis; special immune conditions, for example, paroxysmal nocturnal hemoglobinuria, familial Mediterranean fever; osteoporosis; an ocular disease, age-related macular degeneration; and viral pneumonia in children; and a blood disease, veno-occlusive crisis in sickle cell disease. In oncology, mabs in the form of conjugate molecules also serve as carriers of potent drug toxins to enhance cell killing, for example, brentuximab vedotin. Now, bispecific mabs are developed that bind to two antigens, a cancer cell and an immune cell (e.g., a T-cell lymphocyte) to enhance mab cell-killing activity, for example, Blincyto. The substantial growth in mab products with major new indications is predicated on the specificity of mabs to their cell targets and especially on the process of humanization of the murine antibodies, leading to less side effects and more affinity for the target receptors (more potential desired activity).

Biosimilar mabs are also now developed and commercially available as substitutes for the originator products, for example, infliximab for immune disorders (Remicade versus Inflectra, Ixifi, Renflexis). As of December 2019, 21 biosimilar biotech mab products are commercially available for five different mab molecules and can be used as less expensive alternative product choices. Although the whole biopharma industry is engaged in mab development and marketing, of the 38 companies with a mab marketed product, the most successful companies are Roche (16 products), Amgen (10), Eli Lilly (7), Novartis (7), Johnson & Johnson (6), Sanofi (5), and Merck (5). Some products are co-marketed, a common approach (25 products). Five products have been removed from marketing, mostly related to inadequate marketing activity.

Mabs also were developed as highly specific diagnostic test products; for example, OncoScinct for colorectal and ovarian cancers, carcino-embryonic antigen (CEA) scan for colorectal cancer, ProstaScinct for prostate cancer, MyoScinct for myocardial infarction, Tecnamab Kl for melanoma, Verluma for lung cancer, LeukoScan for osteomyelitis, and Humaspect for colorectal cancer.

chapter nine

Biotechnology products and indications II

The myriad and diversity of biotechnology products besides proteins number over 180 and are discussed in this chapter, including peptides, vaccines, liposomes, oligonucleotides, polymers, cell therapies, and tissue therapies.

Peptides

An alternative to protein therapy is the use of peptides, intermediate-sized molecules, which are comprised of amino acid sequences, but the structures are much less complex. They have fewer amino acids (4 to about 100), few if any disulfide bridges, occasional polypeptide domains, and they are usually nonglycosylated. Their number and uses have grown significantly with over 110 products on the market for over 60 clinical uses in all organ systems, and they are marketed by over 30 companies as shown in Table 9.1. The peptides may be manufactured by recombinant DNA technology or molecularly engineered (synthetic) methods. Molecules can be forms of endogenous (naturally occurring) human peptides, analogs of endogenous peptides, peptides from animal sources, derivatives of natural peptides, novel structural motifs, novel formulations to achieve practical pharmacodynamic activity, or designs with expanded targets for peptides with pharmacodynamics activity. Eleven of these recombinant proteins were identified from animal sources.

The most prevalent use is for diabetes mellitus with 30 products primarily from 4 companies (AstraZeneca, Eli Lilly, Novo Nordisk, and Sanofi). Alterations of peptides via different salt forms, changes in amino acids (additions or deletions), and novel structural molecules have achieved a variety of pharmacologic profiles offering customization of patient therapy. Other hormonal-related uses include growth, osteoporosis, acromegaly, Cushing's syndrome, diabetes insipidus, and labor initiation and control. Cancer uses include prostate, multiple myeloma, and gastroesophageal-pancreatic cancers. Other areas of use are quite broad

Table 9.1 Peptides: [$N = 112 + 5$ in other categories]

Trade name	Nonproprietary name	Company	Usage area
Recombinant – hormones [34 products – 5 uses]:			
Admelog	Insulin lispro	Sanofi	Insulin dep. D.m.
Afrezza	Insulin	Mannkind	Insulin dep. D.m.
Apidra	Insulin glulisine	Sanofi	Insulin dep. D.m.
Basalgar	Insulin glargine	Eli Lilly/Boehr-Ing	Insulin dep. D.m.
Baqsimi	Glucagon	Eli Lilly	Hypoglycemia in D.m.
Bonsity	Teraparatide	Pfenex	Osteoporosis
Exubera[a]	*Insulin*	*Pfizer*	*Insulin dep. D.m.*
Fiasp	Insulin aspart	Novo Nordisk	Insulin dep. D.m.
GlucaGen	Glucagon	Novo Nordisk	Hypoglcemia
Gvoke	Glucagon	Xeris	Hypoglycemia in D.m.
Humalog-R	Insulin lispro	Eli Lilly	Insulin dep. D.m.
Humalog-N	Insulin lispro + protamine	Eli Lilly	Insulin dep. D.m.
Humulin-N	Insulin Protamine	Eli Lilly	Insulin dep. D.m.
Humulin-R	Insulin	Eli Lilly	Insulin dep. D.m.
Increlex	Mecamersin	Tercica	Growth
iPlex	*Somatokine*	*Insmed*	*Growth*
Insulin	*Insulin*	*Bayer*	*Insulin dep. D.m.*
Insulin lispro	Insulin lispro	Eli Lilly	Insulin dep. D.m.
Lantus	Insulin glargine	Sanofi	Insulin dep. D.m.
Levemir	Insulin detemir	Novo Nordisk	Insulin dep. D.m.
Lusdana Nexvue	Insulin glargine	Eli Lilly/Boehr-Ing	Insulin dep. D.m.

(Continued)

Table 9.1 (Continued) Peptides: [$N = 112 + 5$ in other categories]

Trade name	Nonproprietary name	Company	Usage area
Myxredlin	Insulin regular	Baxter	Insulin dep. D.m.
Novolin-L	Insulin	Novo Nordisk	Insulin dep. D.m.
Novolin-N	Insulin isophane	Novo Nordisk	Insulin dep. D.m.
Novolin-R	Insulin	Novo Nordisk	Insulin dep. D.m.
NovoLog	Insulin aspart	Novo Nordisk	Insulin dep. D.m.
NovoMix	Insulin aspart 30% & I-protamine 70%	NovoNordisk	Insulin dep. D.m.
NovoRapid	Insulin aspart	Novo Nordisk	Insulin dep. D.m.
Ryzodeg	Insulin degludec + aspart	Novo Nordisk	Insulin dep. D.m.
Saxenda	Liraglutide	Novo Nordisk	Obesity
Soliqua	Insulin glargine + Lixisenatide	Sanofi	Insulin dep. D.m.
Toujeo	Insulin glargine	Sanofi	Insulin dep. D.m.
Tresiba	Insulin degludec	Novo Nordisk	Insulin dep. D.m.
Victoza	Liraglutide	Novo Nordisk	Diabetes mell. type 2
Xultrophy	Insulin degludec + Liraglutide	Novo Nordisk	Insulin dep. D.m.
Other recombinant peptides [6 products – 6 uses]:			
Fortical	Calcitonin, salmon	Upsher-Smith	Osteoporosis
Gattex	Teduglutide	Takeda	Short bowel synd.
Iprevask	Desirudin	Bausch Health	DVT prophylaxis
Kalbitor	Escallantide	Takeda	Angioedema

(Continued)

Table 9.1 (Continued) Peptides: [N = 112 + 5 in other categories]

Trade name	Nonproprietary name	Company	Usage area
Natrecor	*Nesiritide*	*Johnson & Jhnsn*	*Con. Heart failure*
Refludan	*Lepirudin*	*Bayer*	*Hep. Thrombocytop.*
Molecular engineered (synthetic):			
Hormones [20 products – 13 uses]:			
Adlyxin	Lixisenatide	Sanofi	Diab. Mell. type 2
Antagon	Ganirelix	Merck	Fertilty, LH surge
Bydureon	Exanatide l.a.	AstraZeneca	Diab. Mell. type 2
Bydureon BCise	Exenatide l.a.	AstraZeneca	Diab. Mell. type 2
Byetta	Exenatide	AstraZeneca	Diab. Mell. type 2
Cetrotide	Cetrolix	Merck KGaA	Prevent LH surge
Egrifta	Tesamorelin	Merck KGaA	Lipodystrophy HIV
Forteo	Teriparatide	Eli Lilly	Osteoporosis
Geref	G.H. releasing hormone	Merck KGaA	Growth
Miacalcin	Calcitonin	Novartis	Osteoporosis
Oxytocin	Oxytocin	Par	Labor control
Oxytocin	Oxytocin	Sicor	Labor control
Sandostatin	Octreotide	Novartis	Acromeg., GI tumors
Sandostatin LAR	Octreotide l.a.	Novartis	Acromeg., GI tumors
Somatuline depot	Lanreotide l.a.	Ipsen	Acromeg., GEP tumor
Supprelin LA	Histrelin l.a.	Endo	Precocious puberty
Symlin	Pramlintide	AstraZeneca	Diab. Mell. type 2
Tymlos	Abaloparatide	Radius Health	Osteoporosis
			(Continued)

Table 9.1 (*Continued*) Peptides: [$N = 112 + 5$ in other categories]

Trade name	Nonproprietary name	Company	Usage area
Vantas	Histrelin	Endo	Prostate cancer
Zoladex	Goserelin	AstraZeneca	Prostate ca, Endomet
Other [46 products – 50 uses]:			
Acthar	Tetracosactide	Mallinckrodt	Inflammat. dis. #20
Angiomax	Bivalirudin	Novartis	Coronary angioplasty
Bivalirudin	Bivalirudin x 2	Mylan & Pfizer	Coronary angioplasty
Copaxone	Glatiramir	Teva	Multiple sclerosis
Copaxone	Glatiramir biosim	Mylan	Multiple sclerosis
Cubicin	Daptomycin	Merck	Gram pos. infections
Dalvance	Dalbamycin	Allergan	Gram pos. infections
DDAVP	Desmopressin x 3 forms	Sanofi & Various	D.ins., Noct. polyuria
Eligard	Leuproilde	Talomer Therap.	Prostate cancer
Firazyr	Icantibant	Takeda	Angioedema
Firmagon	Degarelix	Ferring	Prostate cancer
Firvanq	Vancomycin oral	Cutis Pharma	Enterocolitis, MRSA
Fuzeon	Enfuvirtide	Roche	HIV infection
Gengraf	Cyclosporin A	AbbVie	Organ transpl. rejctn
Giapreza	Angiotensin II	La Jolla	Blood pressure shock
Glatopa	Glatiramir	Novartis	Multiple sclerosis
Glatopa l.a.	Glatiramir l.a.	Novartis	Multiple sclerosis
iFactor	Bone peptide	Cerapedics	Cerv. Disect., fusion
Integrelin	Eptifbatide	Merck	Angioplasty, ACS
			(*Continued*)

Table 9.1 (Continued) Peptides: [*N* = 112 + 5 in other categories]

Trade name	Nonproprietary name	Company	Usage area
Istodax	Romidepsin	Celgene/Astellas	T-cell lymphoma
Kyprolis	Carfilzomib	Amgen	Multiple myeloma
Linzess	Linaclotide	Ironwood/AbbVie	Irritable bowel synd.
Lupron	Leuprolide	Takeda/AbbVie	Prostate ca + #4
Lutathera	Octreotide Lu177	Novartis	GEP cancer
Minirin	Desmopressin	Ferring	D.ins., Noct. polyuria
Neoral	Cyclosporin A	Novartis	Organ transpl. rejctn
Noctiva	Desmopressin	AbbVie/Serenity	Nocturnal polyuria
Nocturna	Desmopressin,	Ferring	Nocturnal polyuria
Orbactiv	Oritivancin	Melinta Therap.	Gram Pos. infections
Omontys	*Peginesatide*	*Takeda*	*Anemias*
Parsibiv	Etelcalcetide	Amgen	Hyperparathyroidism
Pitressin	*Vasopressin x 2 forms*	*Kabi & Various*	*Blood pressure stim.*
Plenaxis	*Abarelix*	*Regeneron*	*Prostate cancer*
Prialt	Ziconotide	TerSera Therap.	Pain, severe chronic
Sandimmune	Cyclosporin A	Novartis	Organ transpl. rejctn
Scenesse	Afamelanotide	Amag/Palatin	Hypoactive sexual desire in women
Signifor	Pasreotide	Recordati	Cushing's syndrome
Signifor LAR	Pasreotide l.a.	Recordati	Acromegaly, Cushing's
Stimate	Desmopressin	CSL Behring	D.insi., Noct. polyuria
Surfaxin	*Lucinactant*	*Discovery Labs*	*Resp. distress synd.*

(Continued)

Table 9.1 (Continued) Peptides: [*N* = 112 + 5 in other categories]

Trade name	Nonproprietary name	Company	Usage area
Synarel	Nafarelin	Pfizer	Endometriosis, CPP
Trulance	Plecainide	Synergy	Constipation, IBS
Vancocin	Vancomycin	Eddington	*C. diff., Staph.* MRSA
Vasostrict	Vasopressin	Endo	Blood Pressure stim.
Vasotec	Enalpril	Various	Hypertension
Viadur	*Leuprolide acetate implant*	*Bayer*	*Prostate cancer*
Vylessi	Bremelanotide	Clinuvel	Eryhropoietic proporphyria

In other categories:
Ozempic (Semaglutide), Nplate (Romiplostim), Tanzeum (Albiglutide), Trulicity (Dulaglutide)
[EU: Trelstar (Triptorelin)]

Abbreviations: ACS – acute coronary syndrome, Acromeg. – acromegaly, Boehr-Ing – Boehringer-Ingelheim, Ca – cancer, C. diff. – Clostridium difficile, Cerv. Discect. – cervical discectomy, Co. – company, Con. – congestive, CPP – central precocious puberty, D.insi. – diabetes insipidus, D.m. or Diab. Mell. – diabetes mellitus, Dis – disease, DVT – deep vein thrombosis, Endomet. – endometrial, GEP – gastroesophageal-pancreatic, GI – gastrointestinal, Dep – dependent, Hep – heparin, HIV – human immunodeficiency virus, IBS – irritable bowel syndrome, Jhnsn – Johnson, l.a. or L.A. – long acting, LH – luteinizing hormone, MRSA – methicillin-resistant *Staphylococcus aureus*, Noct. – nocturnal, Pos – positive, Rejctn – rejection, Resp. – respiratory, *Staph.* – *Staphylococcus*, Stim. – stimulation, Synd. – syndrome, Therap. – Therapeutics, Transpl. – transplant.

ᵃ The products in italics have been withdrawn from marketing.

and include blood pressure control, angioplasty, infections (gram positive, HIV), irritable bowel, pain, organ transplants, bone fusion, multiple sclerosis, angioedema, and several inflammatory conditions. There are at least 10 animal sources for peptides, including, for example, leeches (refludan), Gila monster (bydureon), sea snails (ziconatide), and snakes (integrelin).

Vaccine products

Vaccines in biotechnology have become a major focus for the pharmaceutical and biotechnology industries based on the surge in infections that can be prevented, the antigen drift of the viral organisms (changing the offending virus), the huge base of patients needing therapy, and the financial opportunity for companies. Twenty-eight products have been developed for marketing as seen in Table 9.2. Eleven bacteria and viruses are managed with vaccines and involve *Bacillus anthracis*, Dengue virus, Ebola, Hepatitis B, Herpes zoster virus, Human papilloma virus, Influenza, *Neiserria meningiditis*, *Neiserria meningiditis* group B, *Stretococcus pneumoniae*. The vaccines employ recombinant techniques to recreate fractions of a microorganism or virus that retain the activity as an immune stimulant that create immune protection against that infectious species going forward. Hepatitis B vaccines were the first recombinant vaccines created for immunization. Lyme disease had a preventative vaccine available, but the small market caused its withdrawal from sales. Also, fusion molecules are used in vaccines to enhance the immunologic response in patients. Vaccines are also available for the prevention of other serious infections including *Streptococcus pneumoniae* pneumonia (vaccines with 6 serotypes and 13 serotypes) and meningococcal meningitis, both of which are fusion molecules of an immune-stimulant protein and the bacterial polysaccharides. Erbevo vaccine is a fusion protein of Ebola glycoprotein envelope and a derivative of the virus vesicular stomatitis. Another two vaccines with a unique application prevent infection from strains of human papilloma virus, which in turn later prevents the development of female cancers of the cervix and vagina. A new role for vaccines under study is the use as therapeutic products to treat cancer, wherein the vaccine is specific to a cancer type, turns on the patient's immune system against the target cancer, and is used in conjunction with other anticancer treatments. The first commercially available vaccine as a therapeutic agent is sipuleucel-T for prostate cancer.

Liposomes

Liposomes are carrier molecules comprised of lipids most often in spherical molecules with several layers of lipid, and the drug or biological agent is carried within the lipid molecule. The goal is improved product delivery

Table 9.2 Vaccines [N = 28]

Trade name	Nonproprietary name	Company	Usage area
Recombinant: [19 products – 9 uses]			
Bexsero	*N. Meningiditis* Grp B	GlaxoSmithKline	Prevent meningitis B
Cervarix	Papilloma × 2 serotypes	GlaxoSmithKline	Prevent HPV cancer
Comvax	Hepatitis B + *Haemophilus* B	Merck	Prevent Hepatitis B
Dengvaxia	Dengue tetravalent live	Sanofi	Prevent Dengue fever
Engerix-B	Hepatitis B	GlaxoSmithKline	Prevent Hepatitis B
Erbevo	rVSVDelatG-Zebov-GP	Merck	Prevent Ebola
Flubok	Influenza × 3 Antigens	Sanofi	Prevent influenza
Flubok Quad	Influenza × 4 Antigens	Sanofi	Prevent influenza
Gardasil	Papilloma × 4 serotypes	Merck	Prevent HPV cancers
Gardasil-9	Papilloma × 9 serotypes	Merck	Prevent HPV cancers
Heplisav-B	Hepatitis B	Dynavax	Prevent Hepatitis B
Lymerix[a]	*Borrelia burgdorferi protein*	*GlaxoSmithKline*	*Prevent Lyme disease*
Pediarix	Hepatitis B, DPT, Polio	GlaxoSmithKline	Prevent Hepatitis B
Recombivax-HB	Hepatitis B	Merck	Prevent Hepatitis B
Shingrix	Herpes zoster	GlaxoSmithKline	Prevent Shingles
Trumemba	*N. meningiditis* Grp. B	Pfizer	Prevent meningitis B
Twinrix	Hepatitis A & B	GlaxoSmithKline	Prevent Hepatit. A/B
Vaxelis	Diphtheria toxoid, tetanus toxoid, pertussis, poliovirus, *Haemophilus* b conjugate with meningococcal protein, hepatitis B	Merck/Sanofi	Prevent Hepatitis B

(Continued)

Table 9.2 (Continued) Vaccines [N = 28]

Trade name	Nonproprietary name	Company	Usage area
Molecular engineered: [9 products – 6 uses]			
Biothrax	*Bacillus anthracis*	Emergent Biosoltns	Prevent Anthrax
Imlygic	Talimogene laherparepvec	Amgen	Melanoma
Menactra	*N. meningitidis*, diphtheria toxoid	Sanofi	Prevent meningitis
Menveo	*N. meningitidis*, diphtheria cross-reacting material [CRM] conjugate	GlaxoSmithKline	Prevent meningitis
PedVax	*Haemophilus* b, Meningococcal protein conjugate	Merck	Prevent meningitis
Prevnar 7	*S. pneumoniae* 4,6B,9V,14,18C,19F,23F, diphtheria CRM conjugate	Pfizer	Prevent pneumonia
Prevnar 13	*S. pneumoniae* 1,3,4,5,6A,6B,7F,9V,14,18C,19A,19F23F, diphtheria CRM conjugate	Pfizer	Prevent pneumonia
Provenge	Sipuleucel-T	Sanpower	Prostate cancer
Theracys	BCG vaccine	BioChem Pharma	Bladder cancer

Abbreviations: BCG – Bacillus Calmette-Guérin, Biosoltns – Biosolutions, DPT – diphtheria, pertussis, and tetanus, Grp – group, Hepatit. – hepatitis, HPV – human papilloma virus, *N* – *Neisseria*, *S* – *Streptococcus*.

[a] The products noted in column one in italics have been withdrawn from marketing. Italics are used for the names for bacteria.

to target cells, which basically are also lipid sacs, creating improved compatibility in delivery and, hopefully, less systemic toxicity. Five cancer agents (doxorubicin, daunorubicin, vincristine, irinotecan, and cytarabine), one antifungal antibiotic, amphotericin (in three distinct products for several fungal infections), one antibiotic (amikacin) for tubercular disease, and two analgesic products (morphine and bupivacaine) are formulated as liposomes. A few selected serious toxicities are reduced to some extent, in particular, cardiac damage with the doxorubicin product and kidney damage with the amphotericin product (Table 9.3).

Oligonucleotides

Inhibition of aberrant mRNA and replacing nonfunctional mutated genes as causes of serious disease has been a very major development challenge especially regarding and requiring intracellular delivery and optimal cell targeting. Improvement in oligonucleotide delivery has resulted in very promising products in the last few years. Eleven products have been approved for patient care, see Table 9.4, The first antisense molecule (fomivirsen) was used to treat Cytomegalovirus (CMV) retinitis in HIV patients, but it has

Table 9.3 Liposomes-liposome [$N = 13$ products – 14 uses]

Trade name	Nonproprietary name	Company	Usage area
Abelcet	Amphotericin B-liposome	Leadiant Biosciences	Aspergillosis
AmBisome	Amphotericin B-liposome	Astellas	Fungal infections
Amphotec	Amphotericin B-Liposome	Kadmon	Cryptococcal mening.
Arikayce	Amikacin-liposome	Insmed	Avian tuberculosis
DaunoXome	Daunorubicin-liposome	Diatos	Kaposi's sarcoma +2
DepoDur	Morphine-liposome	Rafael	Post-op. pain
DepoCyt	Cytarabine-liposome	Leadiant Biosciences	Meningitis lymphad
Doxil	Doxorubicin-liposome	Ipsen	Kaposi's sarcoma +2
Exparel	Bupivacaine-liposome	Pacira	Pain, Nerve block
LipoDox	Doxorubicin-liposome	Caracao	Kaposi's sarcoma +2
Marqibo	Vincristine-liposome	Acrotech BioPharma	Lymphocytic leukem.
Onivyde	Irinotecan-liposome	Ipsen	Pancreatic cancer
Vyxeos	Daunomycin+liposome	Jazz	Myeloid leukemia

Note: The + sign in the usase area indicates additional uses for the product.

Abbreviations: Post-op. – post-operative, leukem. – leukemia, mening. – meningitis, lymphad. – lymphadenopathy.

Table 9.4 Oligonucleotide/gene therapies: [N = 11 products + 2 in EU – 10 uses]

Trade name	Nonproprietary name	Company	Usage area
Defitelio	Defibrotide	Jazz	Hepatic veno-occlus.
Exondys 51	Eteplirsen	Sarepta Therap.	Duch. musc. dys.51
Givlaari	Givosiran	Alnylam	Hepatic porphyria
Kynamro	Mipomersen	Kastle Therap.	Hypercholesterol.
Luxturna[a]	Voretigene neparvovec-rzyl	Roche	Retinal dystrophy
Macugen	Pegaptanib	Gilead	Macular degenerat.
Onpattro	Patisiran	Alnylam	FHT amyloidosis
Spinraza	Nusinersin	Biogen/Ionis	Spinal muscle atrop.
Tegsedi	Inotersen	Ionis/Akcea	FHT amyloidosis
Vitravene[a]	*Fomivirsen*	*Ciba*	*CMV retinitis*
Vyondys 53	Golodirsen	Sarepta Therap.	Duch. musc. dys.53
Zolgensma[a]	Onasemnogene abeparvovev-xioi	Roche	Spinal muscle atrop.

[E.U.: Glybera[a] (Alipogene tiparvovec) - Lipoprotein lipase deficiency, Strimvelis[a] (SCID gene) -ADA-SCID]

Abbreviations: Atrop – atrophy, ADA – adenine deaminase deficiency, CMV – cytomegalic virus, Degenerat – degeneration, Duch. musc. dys. – Duchenne's muscular dystrophy, FHT – familial hereditary transthyretin, Hypercholesterol. – hypercholesterolemia, Musc – muscular, Therap – therapeutics, Veno-occlus. – veno-occlusive crisis, SCID – severe combined immunodeficiency.

[a] Gene therapies.

been removed from the market because of low usage related to improved anti-HIV treatments and little CMV retinitis. Difficult administration (multiple intra-vitreal injections) limits Macugen use for its ophthalmologic indication. In the rare disease area, severe familial hypercholesterolemia, transthyretin amyloidosis and Duchenne's muscle atrophy now have treatments. Gene therapies with curative potential finally have become available for Duchenne's muscular dystrophy and spinal muscle atrophy.

Tissue and cell engineering products

Cell and tissues therapy products are found in Table 9.5. Surgery, especially gastrointestinal and orthopedic (back) locations, can lead to complications where abnormal connections called adhesions can occur between tissues. They can be persistent and quite painful after surgery and often require a second surgery to eliminate them. Hyaluronic acid products in the form of gels and films are available to prevent them by placement between tissues during surgery. The products are biodegradable *in situ* to nontoxic substances. Also, tissue damage occurs in various diseases

Table 9.5 Tissue/cell engineering [N = 18 products – 20 uses]

Trade name	Nonproprietary name	Company	Usage area
Chondrocytes: [4]			
Carticel	Chondrocytes	Vericel	Cartilage damage
Cartiform	Chondrocytes, extracellular matrix, collagen	Athrex	Cartilage repair
Fortaflex	Collagen scaffold	Organogenesis	Rotator cuff repair
MACI	Chondrocytes, cell matrix	Vericel	Knee cartilage
Fibroblasts:			
LaViv	Azficel-T	Fibrocell Science	Nasolabial folds
Mesenchymal cells: [2]			
Bio4	Mesenchymal cells, osteoprogenitor cells, osteoblasts, angiogenic g. f.	Stryker	Bone repair
Ovation[a]	*Mesenchymal cells*	*Osiris Therap.*	*Surgical wounds*
Tissue/skin cells: [9]			
Apligraf	Skin substitute – epithelial cells	Organogenesis	Wounds, Foot ulcers
Dermagraft	Skin substitute – fibroblast cells	Organogenesis	Foot ulcers
Epicel	Skin substitute – keratinocytes, fibroblasts	Vericel	Burns, skin repair
Gintuit	Skin substitute – keratinocytes, fibroblasts, collagen	Organogenesis	Mucogingival repair
Grafix	Skin substitute – epithelial, fibroblast, mesenchymal cells, collagen	Smith & Nephew	Wound repair
Integra matrix	Dermal layer cells	Integra Life Sciences	Burns, Scars
Integra Template	Dermal layer cells	Integra Life Sciences	Burns, Scars

(Continued)

Table 9.5 (Continued) Tissue/cell engineering [N = 18 products – 20 uses]

Trade name	Nonproprietary name	Company	Usage area
Orcel	Skin substitute – epidermal cells, keratinocytes, fibroblasts, collagen,	Forticell Bioscience	Epiderm. Bullosa
TranCyte	Neonatal foreskin fibroblasts	AbbVie	Burns, skin repair
T-lymphocyte cells: [2]			
Kymriah	Tisgenleleucel	Novartis	B-cell leuk. & lymph.
Yescarta	Axicabtagene ciloleucel	Gilead	B-cell lymphomas +3
Products in other categories (vaccines):			
Provenge (Sipuleucel-T), Therasys (BCG)			

Note: The + sign indicates the number of uses for a product.

Abbreviations: BCG – Bacillus Calmette-Guérin, Epiderm. – epidermolysis, g.f. – growth factor, Leuk. – leukemia, Lymph. – lymphocytic cancers.

[a] The products in italics have been withdrawn from marketing.

where the tissue is accessible for replacement, for example, skin ulcers from diabetes, or pressure. Skin grafting can be done with exogenously engineered skin products. Repair of wounds and burns can be aided with these grafts. Tissue damage is becoming a greater problem as the population ages, and tissues tend to break down more over time in older populations, for example, osteoarthritis of the knees or facial wrinkling. Knee pain can be relieved and wrinkling reduced with hyaluronic acid products administered directly into tissues, and even chondrocytes can be replaced in the knee. Bone fractures can be mended more rapidly through enhanced processes with biological products or devices that contain bone morphogenic growth proteins (BMPs). Facial lipodystrophy in HIV patients can be reduced with a biological product in a form of lactic acid.

Another technique to treat disease is based on obtaining healthy cells from a specific tissue, selecting out a specific subset of cells with certain desirable properties, and enhancing the activity of these cells through *ex vivo* manipulation. We then return these specifically selected, enhanced, and activated cells to patients whose cells are not sufficiently functional, thereby ameliorating a disease. Currently, chondrocytes responsible for cartilage production are taken from a patient's knee that has serious damage and is repairing poorly. These chondrocytes are manipulated *ex vivo* and returned to the patient to normalize cartilage production. Bone marrow progenitor cells are collected from peripheral blood, bone marrow cells, or cord blood, and the progenitor cells with greatest regenerative potential (CD34 cells) are selected through various cell-tagging processes. Following life-threatening high-dose chemotherapy in a cancer patient, which destroys almost all the bone marrow, these selected progenitor cells are administered to the patient to accelerate regeneration of bone marrow and white blood cell production, thereby preventing infections. Foreskins of newborns are collected and placed in a three-dimensional construct that permits growth of dermis and epidermis. This construct is used later in venous leg ulcers to accelerate wound healing, thereby reducing health care needs. In tissue engineering, generally, we need tissue or cells from patients or normal donors through biopsy, a process for *ex vivo* cell expansion, a scaffold on which to grow the cells, and a bioreactor system in which the new tissue can grow to its full size with normal structure and functional capacity and devoid of toxicity, immunogenicity, or fibrosis.

Cell manipulations have advanced in the field of immune-oncology, wherein immune cells such as T-lymphocytes that normally are stimulated naturally to fight cancers are substantially enhanced *ex vivo* in their immune cancer fighting function. T-cells are taken from individual patients and by *ex vivo* genetic and cell manipulations in a laboratory are newly altered to be multifaceted as targeting receptors, so that the T-cells' immune reaction versus the targeted cancer yields major substantial cell killing activity.

Polymers and other biotech products

Table 9.6 lists a variety of different product types for a variety of indications that do not fit in the previous categories in Chapter 9. A biological carrier (wafer) is a polymer-drug combination, and is used to facilitate the administration of the cancer drug, BCNU-carmustine, during brain

Table 9.6 Polymers and other products [$N = 3 + 7$ in other categories – 10 uses]

Trade name	Nonproprietary name	Company	Usage area
Gliadel [P]	Carmustine wafer	Arbor	Glioblastoma
Opaxio [P]	Polyglumex-Paclitaxel	Novartis	Breast cancer
Sculptra [P]	Poly-L-lactic acid +	Galderma	Facial lipoatrophy +2

Polymers in other categories:

Bydureon LAR (Exenatide LAR), Eligard (Luprorelin polymer), Lupron Depot (Leuprorelin + microspheres), Sandostatin LAR (Octreotide LAR + microspheres), Signifor LAR (Pasreotide + microspheres), Supprelin LA (Histrelin implant), Vantas (Histrelin implant)

Other products:

Ruconest	Conestal alfa	Pharming	Angioedema
Endari	l-glutamine	Emmaus medical	Sickle cell disease

Abbreviation: LA or LAR – long-acting release.

surgery for the tumors. Such products also further represent the expanding breadth of indications and new types of molecules for biotechnology products. Often the polymer formulation results in longer-acting products. Finally, a couple products are listed above and are unique within themselves, conestal alfa, created in transgenic rabbits, and -glutamine, which is a single amino acid.

Biological products (blood derivatives, tissue extracts, vaccines)

Many biological products (221 products for 121 indications) have been and are developed by traditional methods, distinct from biotech products from recombinant proteins and monoclonal antibodies from hybridoma technology. Commonly, they are extracts or derivatives of the actual protein from the human body, most often from blood or specialized animal sources, plus traditionally developed vaccines employing egg cultures of infectious agents. This traditional approach continues today for blood derivatives, vaccines, tissue or cell extracts, and animal extracts for therapeutic use. Since this book addresses products particularly derived from biotechnology, we will only outline these products in this book. For human blood–derived products, 80 products are marketed for 37 indications. Albumin is obtained for cardiovascular volume conditions. Immunoglobulins (Igs) extracted from blood are available for immunodeficiency conditions, viral infections (hepatitis B and vaccinia), hemolytic anemia in newborns, and idiopathic thrombocytopenic purpura. Anti-hemophilia products are still derived from blood. An anti-globulin is produced for kidney transplant rejection. Several newer products have been developed for hereditary angioedema. Fibrin sealants are created to stop intra-surgical bleeding. Progenitor blood cells are obtained for stem cell transplants used with cancer chemotherapy. Vaccines (66 products) include prevention of viral and bacterial infections (26 organisms). Vaccines are formulated as single organism/infection products and also in various combinations. Tissue and cell extracts include 69 products for 50 indications. Animal blood extracts serve primarily as antidotes and number 8 for 7 indications. Table 9.7 lists the many types of biological products marketed in the United States.

Table 9.7 Traditional biologic products

Blood derivatives:	Vaccines: (26 Organisms)
Albumin	Adenovirus-4,7
Anti-hemophilia factors	Anthrax
Alpha-1 antitrypsin factors	BCG
Fibrin sealants	Cholera
Immunoglobulins	Dengue
Plasma proteins	Diphtheria
Progenitor cell transplants	Encephalitis, Japanese
	Haemophilus b
Tissue/cell extracts:	Hepatitis A
Collagen	Hepatitis B
Dermal matrix	Herpes zoster
Enzymes	Influenza A & B strains
Extracellular matrix	Measles
Heparin & related products	Meningococcus C&Y
Hormones	Meningococcus B
Hyaluronidase	Mumps
Interferons	Pertussis
Pollen extracts	Pneumococcus
Surfactants	Polio
Snake venoms	Rabies
	Rotavirus
Animal blood extracts:	Rubella
Antithymocyte	Smallpox
Black widow spider antidote	Tetanus
Botulinum toxin antidote	Varicella
CroFab	Yellow fever
DigiFab	
Scorpion antidote	

Abbreviation: BCG – Bacillus Calmette-Guérin.

chapter ten

Biotechnology industry

The biotechnology area has become a cornerstone in novel product development for advanced patient care for both biotechnology and pharmaceutical companies, collectively called "biopharma" companies. About one-third of all new product approvals by all companies now are biotech products annually. By the late 1970s/1980s, the science of biotechnology had evolved sufficiently for the creation of biotech companies focused on developing, manufacturing, and then marketing biotech products. In the initial first wave (1971–1982), about 20 biotech companies were founded and led by Biogen, Cetus, Chiron, Genentech, Genzyme, and Amgen.

Biotech companies

About 100 biotech companies over the last 50 years, from 1970 to 2018 and across the world, can be considered historically the leading companies in biotechnology. This top 100 status is based on one or more of the four following factors: (1) performing the research, manufacturing, and marketing of a biotech product on their own without pharma company engagement, and/or (2) contributing principally to a significant product's discovery and development that was marketed later by someone else, and/or (3) dedicating substantially in biotech research funding up to at least $100 million for at least 4 years, and/or (4) being acquired by a biopharma company at a substantial cost because of their science platform was novel and product potential was high at that time. See Table 10.1 for the list of top 100 companies over 1970 to 2018. The top 20 companies are listed via bold print also over these 50 years of biotechnology companies' existence, based on number and importance of products developed and marketed, plus extent of revenues. The very top-line, worldwide companies' number nine who achieved exceptional research and marketing, including Amgen (USA), Biogen (USA), Chugai (Japan), Genentech (USA), Genzyme (USA), Novo Nordisk (Denmark), Regeneron (USA), Roche (Switzerland), and Serono (Italy). Furthermore, biotech success is well evidenced by 43 of these 100 biotech companies having been acquired by usually a pharma or another biotech, as noted in italics in Table 10.1, an amazing 43%. The United States has led the biotech industry in products developed, their patient care impacts, and marketing.

Table 10.1 Top 100 biotech companies over 1980 to 2018

Abgenix[a]	Actelion	*Affymax*	Alexion
Alkermes	*Allergan*	Alnylam	*Alza*
Amgen[b]	*Amylin*	*AxeVis*	Beigene
Biogen[b]	**BioMarin**	Bluebird Bio	BTG Internat'l
Cambridge AT	Celldex	*CellTech*	*Centocor*
Cetus	***Chiron***	***Chugai***[b]	Clovis Oncology
Cor	*Corixa*	**CSL Behring**	CTI Biopharma
Cubist	Daiichi-Sankyo	*Dyax*	Dynavax
Elan	Emergent Biosol.	Enzon	Eyetech
Exelis	Ferring	Galapagos	***Genentech***[b]
Genetics Institute	GenMab	***Genzyme***[b]	Glyart
Halozyme	*Human Genome Sci.*	*Idec*	Ilex
Imclone	***Immunex***	ImmunoGen	Incyte
Insmed	Ipsen	**Ionis (Isis)**	Ironwood
Jazz	*Juno Therapeutics*	*Kite Pharma*	**Kyowa-Hakko-Kirin**
Ligand	*Liposome*	Mannkind	*Medarex*
MedImmune	*Medivation*	Merrimack	Merz
Micromet	*Millenium*	Morphosys	Nektar
NeXstar	**Novo Nordisk**[b]	*NPS*	*Onyx*
Organogenesis	Organon	**PDL Biopharma**	*Pharmacyclics*
Portola	*Protalix*	*Protein Sciences*	Puma Oncology
Radius Health	**Regeneron**[b]	**Roche**[b]	Sarepta Therapeutics
Scios	**Seattle Genetics**	*Sequus*	***Serono***[b]
Shire	SOBI	*Spark Therapeutics*	Tanox
Theratechnologies	ThromboGenics	*TKT*	UCB
Ultragenx	United Therap.	***Wyeth***	Xencor
Xoma			

Abbreviations: Biosol. – Biosolutions, Cambridge AT – Cambridge Antibody Technologies, Sci – Sciences, Therap. – Therapeutics, TKT – TransKaryotic Technology.

[a] The companies in italics have been acquired by a pharma or another biotech.
[b] The top 20 companies are indicated by bold print. Some companies are both, which is indicated by boldprint and italics.

Furthermore, the top 25 worldwide pharma companies (drug-focused) are now all involved in biotech product research and marketing. The early major players in the 1980s included Eli Lilly, Johnson & Johnson, Novo Nordisk, and Roche. Pharma companies started their biotech engagement with in-licensing biotech products from biotech companies in the 1980s, followed by biotech company acquisitions and then full-fledged internal biotech research units; now all three strategies are utilized. Research and development spending reached an estimated $65 billion in 2018 by only the top 54 biotech and top 13 pharma companies. The companies engaged in the marketing of biotech products expanded dramatically from about 30 in the 1990s to over 100 in 2018. Additionally, in 2018, about 680 companies worldwide had about 1,900 biotech products in clinical research in the United States and Europe in patients for nearly 300 medical conditions. Furthermore, over 4,000 biotech companies worldwide employed over 175,000 staff in the 2015–2018 timeframe. Some sources tabulate as many as 10,000 companies worldwide. Revenues from biotech products have risen dramatically over the last 30 years to approximately $264 billion in 2018 from $75 billion in 2010 and earlier about only $5 billion in 1990 collectively in the United States, Canada, Europe, Japan, and Australia. The emerging countries in medical care – especially China, and also India, Brazil, Russia, Vietnam, South Korea, and others – will be adding substantially to biotech research, product development, and product sales going forward. China has already established a major presence in biotech start-up companies, for example, Beigene, Ascletis Pharma, Cstone, and Innovent, raising well over $100 million to $900 million in their initial public offerings. A representative geographic distribution of companies in the United States is shown in Figure 10.1; California and Massachusetts continue to be the leading areas of biotechnology research and company creation, aligned generally with the concentration of medical universities engaged in biotechnology research. The East Coast corridor from Connecticut to North Carolina for pharma companies has also observed many biotech companies being created.

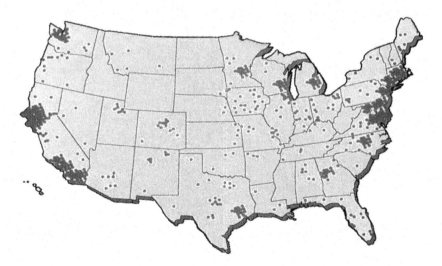

Figure 10.1 Biotech companies – locations in the United States. (Reprinted with permission from Taylor and Francis, Encyclopedia of Pharmaceutical Technology.)

Challenges of a biopharma company

All biotech companies, and drug companies as well, have a myriad of science, heath care, and business challenges impacting their organizations and operations that must be addressed in order to achieve success in developing-manufacturing-marketing medically needed products with reasonable profitability. Figure 10.2 presents 12 such challenges. A *"revolution in biotechnology"* is occurring for targets, mechanisms, treatment alternatives, and molecular structures for disease mitigation and cure in, for example, newer technologies in immune-oncology, gene editing, molecular engineering in 2015–2020 timeframe. Diseases currently treatable with biotech products number over 390, and diseases are being subcategorized at multiple levels with genetics and stages of disease, requiring different product choices; such that the *"complexity of diseases"* has increased dramatically over last 20–30 years. With over 440 biotech products marketed, *"competition"* among products for utilization can be fierce for prescriptions and reasonable market share. For example, in psoriasis, 20 biotech products (11 unique molecules) are available for its chronic therapy, each with quite favorable disease-mitigating actions.

"Customers" are many in number and variety and are always an ever-evolving audience for biopharma industry. An extensive list of groups (customers) exist with different demands, information needs, and support needs: (1) patients and their support groups; (2) health care professionals

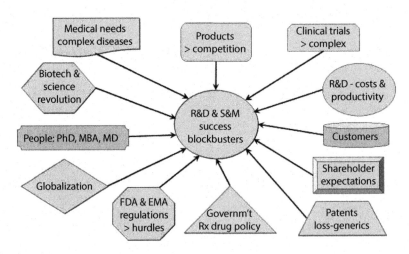

Figure 10.2 Challenges for biopharma companies.

Abbreviations: EMA – European Medicines Agency, FDA – U.S. Food and Drug Administration, MBA – Masters in Business Administration, MD – Medical Doctor, PhD – Doctor of Philosophy, R&D – Research and Development, Rx – Prescription, S&M – Sales and Marketing. (Reprinted with permission from Amgen Inc.)

(physicians – prescribers and pharmacists – the gatekeepers, and their professional societies); (3) health care systems with their access controls (hospitals, prescription benefit managers, managed care organizations); (4) payers, such as insurance companies, Medicare, Medicaid, and the Veterans Administration; (5) investor relations (shareholders and other investors); (6) public relations areas (the public and press); (7) policy-makers (congressmen), and (8) even publishers of medical journals with research design standards for publications. The many *"government"* agencies and legislative bodies present challenges at many levels, for example, Medicare, Medicaid, Pentagon, Health & Human Services, Congress, State legislators and administrators, who create policy-making and legislation regarding drug access, research funding, and payments for products. The U.S. Food and Drug Administration (FDA) and European Medicines Agency (EMA) with their *"regulations,"* *guidances*, and oversight are the primary gatekeepers for product availability (approvals based on safety, efficacy, and quality of products), usage guidelines, safety monitoring, and standards for research-manufacturing-marketing. *"Patents"* from government patent offices offer protection to a company and their products versus competitive products regarding a products' availability and usage during the patent's lifespan, but patents will expire, eventually leading to copycat molecules and competition. *"Globalization"* impacts research and development and marketing and sales with differential international needs (government, disease, cultures, prevalence, and presentation) in the four major markets (the United States, Europe, Asia, and developing countries), and especially worldwide coordination. *"Clinical trials"* provide the data that drive the companies' success to meet both research requirements and marketing programs to ensure greater and appropriate product usage. Growing demands in the quantity and complexity of the clinical trials are never-ending challenges. *"Shareholders"* invest in companies and need demonstration of a company's excellence in strategies, efficient operations, new products generation, and near- and long-term profitability. The *"people"* resource (employees) is a cornerstone of success at any company, and the professional staff of scientists and marketers (PhDs, MDs, MBAs) must be recruited, trained, challenged, excited, rewarded, and retained.

Culture of a biotech company

The origin of a biotech company very often starts with a discovery by a university scientist that may be a new and different pathogenesis (mechanism) for a disease that directly relates to disease progression and its beneficial cessation or stimulation, or the discovery of a novel technology platform to create novel molecules, or a novel molecule (product candidate) with therapeutic and financial opportunity. A new "spin-out" company from the university is formed with a heavy research focus, requiring research and operational funding from outside the university. The culture over their first few years of existence of these early stage biotech companies is hallmarked by nine common traits: (1) small size (10 to a few hundred staff), (2) cutting-edge research focus, (3) academic style environment, (4) open and challenging communication style, (5) team-based work, (6) open iterative free-wheeling decision process (best ideas predominate), (7) flexibility and nimbleness, (8) budgeting via venture capital rounds and government agency grants, and (9) management team and board of directors of mostly scientists. As the science advances and products come into focus, the culture and operational needs evolve as well as follows. The traits of openness, challenging, teams, nimbleness, and cutting-edge need to be sustained and reinforced, but more planning and management oversight/process (projects-products-portfolio) are required, with more product focus on market expectations, as well as disease mitigation. Management brings in industry-experienced people (research, regulatory, and marketing) to help guide the organization through new stages and operational requirements. More focus is on the lead products along with out-licensing of non-target work. Budgeting evolves as well, over time, as research and development (R&D) progresses, which is discussed below.

Financing a biotech company

The source of funds for R&D and operations for a biotech company includes 10 different sources as the company grows and evolves, especially in the science and marketing areas, as outlined in Table 10.2. The corporate goals during this evolution are continued R&D successes moving from a private company to a public company, and eventually to becoming a "fully independent pharmaceutical company" (FIPCO) with the three core functions of R&D, sales and marketing (S&M), and manufacturing functions. Alternative company outcomes are remaining an R&D company with out-licensing of their novel molecules or acquisition by another larger biopharma company. Approximate funding amounts also are suggested for each situation based on typical history. Companies almost always start as private companies with relatively low funding needs, $10–50 million annually. Venture capital companies most often are the initial source of funds to fulfill early space needs (labs and offices), hire the first group of scientists, and initiate the research programs. Partnerships with other companies is by far the largest source of funding, often 40%–50% of the total. Almost all companies utilize most of these various funding sources as they evolve and grow.

Figure 10.3 presents the overall funding in 2018 for public biotech companies in the standard six categories, compiled annually by *Nature Biotechnology*; 460 public companies (about 50% in United States) collected $161 billion. This overall funding level varies greatly year to year based

Table 10.2 Biotech company financing

---

1. Venture capital: *"IDEA"* stage – First funding round at the founding of a private company [a few million dollars],
   - CEO/CSO, board of directors, and science advisory board formed and hired and core small R&D staff with laboratory space primary needed.
   - Up to five subsequent VC funding rounds B through F, every few years, as needed to fund the science and product advancement with added and new staffing and space as needs evolve [$50–500 M].
2. "Angel" investor: Wealthy individual seeds the money for the start-up company [$.5-5 M].
3. Government Agency Grants: *"IDEA"* & early stage research – SBIR [$.2–2 million], NIH and NCI. [$0.1 to several million], ODD/FDA [ave. about $.3 million, 85 grants/year].
4. "Partnerships" with biopharma company: Sharing of *"Products"*
   - Cost sharing for costs of R&D & operations & biotech build-out, along with later sales sharing or royalties on sales.
   - Payments to a biotech company occur at significant milestones (e.g., R&D events – Phase 1, 2, or 3 initiation, publications, regulatory filings) [$10–950 million].
5. "Outlicense": Sell-off a *"product"* in the portfolio, while focusing on a lead molecule, or sell access to their novel technology platform [$.5–5 Billion].
6. "IPO": Initial (first) stock offering as the *"Science and/or Product"* advances significantly with external recognition of the existing products and/or science and business opportunity [$25–900 million].
7. Follow-on stock offerings, as Science/Product advances even further [$25–250 million].
8. Patents: "Court cases": Your "Science" wins as the innovation versus a rival company in a court case, limiting competition and establishing novelty of your science, affording more follow-on VC investments or stock sales conducted.
9. Debt: Bank loans obtained for "special targeted needs"; e.g., BLA preparations, Build Plant for Manufacturing, Company or product acquisitions [$.25–50 B].
10. "Product Approval" & "Sales" of products.

Abbreviations: ave – average, B – billion, BLA – Biologics License application, CEO – Chief Executive Officer, CSO – Chief Science Officer, FDA – U.S. Food and Drug Administration, IPO – Initial Public Offering, M – million, NCI – National Cancer Institutes, NIH – National Institutes of Health, ODD – Orphan Drug Designation, R&D – Research and Development, SBIR – Small Business Innovation Research, VC – Venture Capital.

on, at least, the overall economy, risk aversion, significant new discoveries (technologies or products), disease advances, sales picture, and government intervention (desired and undesired). The competition for resources in the investment community (versus other types of investments) and between companies is quite fierce.

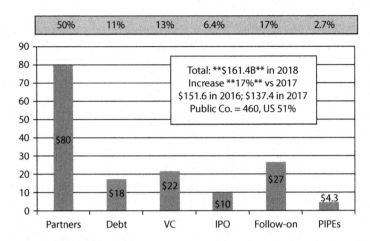

Figure 10.3 Biotech company financing example for 2018.

Abbreviations: B – billion, Co. – Companies, IPO – Initial Public Offering, PIPE – Private Investment in Public Equity, US – United States, VC – Venture Capital. (with permission from *Nature Biotechnology*.)

A biopharma company structure

A FIPCO for both pharma (drugs) and biotechnology (biologicals) includes all the standard and necessary functional areas to create and market a product, starting with the five major functions, that is, Research, Development, Marketing, Sales, and Manufacturing plus eight additional standard support divisions, that is, Medical Affairs, Global Operations, Human Resources, Legal, Regulatory, Finance, Investor Relations, and Public Relations. The next six figures display each of the major divisions within a company and summarizes their primary roles and activities.

The **Research Division** (Figure 10.4) creates product candidates to advance onto Development (clinical research) involving nine steps, in order of evolution from an idea to a product "candidate": (1) starting with study of disease pathogenesis; (2) identifying new *"targets"* that are primary to the disease (identify and validate targets); (3) finding or designing molecules among the many biotech technologies, such as peptides, proteins, monoclonal antibodies, cell therapy, gene therapy or others, that have positive *"hits,"* that is, positive activity on the disease targets; (4) ferreting out the lead molecules with the best hit (impact) on the targets (identify and validate hits to create the *"lead"* molecules); (5) performing preclinical research in animals for molecule's *"pharmacodynamics,"* its profile regarding disease activity and toxicity of the molecule in the disease being studied, establishing its *"proof of principle"*; (6) creating a *"formulation"* of the molecule, for example, tablet or injection or alternative with its various adjunctive ingredients (maintaining its activity, no added toxicity, optimal pharmacokinetic profile, plus extended shelf life) to be best used in future patients with the disease; (7) studying the *"metabolism"* and pharmacokinetic profile of the molecule in animals in liver, kidneys, and blood; (8) examining any *"genomic markers"* that may predict, enhance, or limit a molecule's use; and (9) the end result is a *"product candidate"* with an optimal chance to potentially function with quality, efficacy, and safety in humans. The product candidate is then handed over to the Development division for human trials. Overall ideal goals in discovery are manifold; more candidates with new advantageous properties, more activity, novel mechanisms of action, target specificity, and less toxicity, and all done more rapidly at less cost (Figure 10.4).

The **Development Division** takes on the product candidate from the research division to perform the clinical trials with patients with the target disease and establish its efficacy and safety (Figure 10.5). Clinical trials in humans progress from Phase 1 (with normal subjects for product activity and safety), to Phase 2 (in patients with target disease to document disease improvement, reasonable dosing schemes, safety), and metabolism (absorption-distribution-metabolism-elimination studies in normal subjects and patients), to Phase 3 (pivotal studies in target patients to establish safety and efficacy sufficient for product approval for marketing). The biostatistics group summarizes and analyses the data from the clinical trials for statistical reports. Project management coordinates and tracks all the studies and helps keep projects on time and on target in support of everyone else. Quality assurance is an internal audit function to ensure all the projects and reports are of the highest quality to ensure product approval by regulatory authorities. Resources may be limited as a candidate moves through the phases of research, creating the need to outsource a function

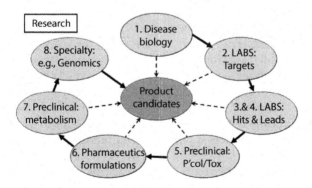

Figure 10.4 Research division of a FIPCO.

Abbreviations: FIPCO – fully independent pharmaceutical company, P'col – Pharmacology, Tox – Toxicology. (Reprinted with permission from Amgen Inc.)

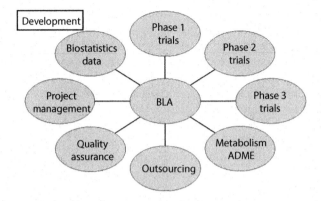

Figure 10.5 Development division of FIPCO.

Abbreviations: ADME – absorption, distribution, metabolism, and elimination studies, BLA – Biologics License Application, FIPCO – fully independent pharmaceutical company. (Reprinted with permission from Amgen Inc.)

to a vendor with the requisite expertise, usually called a contract research organization (CRO). Clinical data from the patient trials with the accompanying clinical trial reports are the outcomes from the Development Division.

The **Manufacturing Division** (Figure 10.6) in biotech makes the products through recombinant technology or molecular engineering for use both during all the R&D steps and for marketing for general use by the medical community in patients. Manufacturing usually includes the physical plant and all the machinery, formulation preparation of the products, assurance and testing for quality of products, package engineering for best available container systems and its labeling, continual process engineering to produce the best product at the least cost, inventory control and distribution for the product to timely meet demands according to the sales forecasts for the country of origin and the rest of the world. Manufacturing evolves from small needs for the biotech molecules for laboratory research, all the way through the very large scale-up for product needs after its marketing approval. Worldwide needs must be met and coordinated for the United States, Europe, Asia, and the rest of the world (Figure 10.6).

The **Marketing Division** (Figure 10.7) entails both marketing and sales functions for products approved for marketing by regulatory authorities. Marketing creates the plans and promotional materials employed by the salesforce in fostering product usage by health care professionals. Plans of action ("POAs") are created in detail often quarterly and annually to foster salesforce actions and product sales. Marketing roles also include creating promotional materials (advertisements) in medical journals and other venues, educational programming for health care

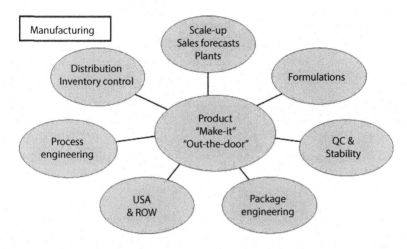

Figure 10.6 Manufacturing division of a FIPCO.

Abbreviations: FIPCO – fully independent pharmaceutical company, QC – quality control, ROW – rest of the world, USA – United States of America. (Reprinted with permission from Amgen Inc.)

professionals, direct-to-consumer advertising as deemed necessary, professional relations (with medical and pharmaceutical societies), R&D liaison & coordination (disease and product information), market research (disease, patient, and practitioner needs, marketplace barriers, competitive products and companies), sales forecasting (potential sales projections by disease, by product, by regions and districts, by sales representative, over time by POA, annually, for next 5 years). Key functions with the salesforce include recruiting, hiring, training, and compensation of sales people; educating them regarding new products, new information, new sales materials; and, of course, their selling activity with customers (health care professionals, health care systems, and payers). In each POA and promotional plan, six steps are utilized by marketing staff in each sales cycle: (1) strategic planning (the guide and goals), then (2) program development (the educational, informational, and sales materials), (3) budget work-up (costs in materials and people), (4) roll-out (via salesforce), (5) measuring success (product sales, audience acceptance), and (6) assess for possible use on next POA (Figure 10.7).

The **Global Division** (Figure 10.8) for the most part addresses the rest of the world outside of the primary market area for a product for the major internal functions of R&D, S&M, and manufacturing. External issues are considered to be of paramount need in successful execution by the global division, such as the language and medical culture of other countries impacting the content and use of educational and informational materials and plans of action. Approvals for product marketing have to be repeated

Figure 10.7 Marketing and sales division of a FIPCO creates the sales plans.

Abbreviations: D-T-C – direct-to-consumer, FIPCO – fully independent pharmaceutical company, POA – plan of action, R&D – research and development. (Reprinted with permission from Amgen Inc.)

in all countries across the world; the European Union has a pan-European system for member countries for the medical regulatory approvals. However, most countries have socialized medicine with the government as the single payer and purchaser of products. Therefore, product prices must be negotiated with every country individually. Finally, the internal issues of R&D, S&M and manufacturing require coordination and integration of all global efforts across the world, which are critical to success of any multinational company (Figure 10.8).

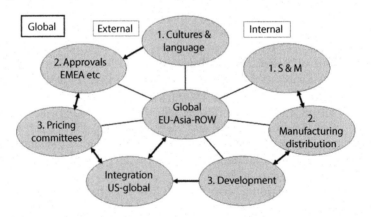

Figure 10.8 Global division of a FIPCO.

Abbreviations: EMEA – European Medications Evaluation Agency, EU – European Union, FIPCO – fully independent pharmaceutical company, ROW – rest of the world, S&M – sales and marketing, US – United States. (Reprinted with permission from Amgen Inc.)

Table 10.3 Support functions/departments in a FIPCO

• **Regulatory** • All communications with regulatory authorities • Applications: NDA, BLAs, Supplements • Audits	• **Finance** • Annual/quarterly reports • Accounts receivable • Accounts payable • Budgeting oversight
• **Legal** • Patents • Contracts [customers, licensing, employees] • Law suits • Merger & acquisitions	• **Investor relations** • Shareholders & investment community • **Public relations** • Press & public
• **Medical affairs** • Medical oversight of marketing • HCP communications for products	• **Human resources** • Recruiting & training • Compensation • Promotions & retention

Source: Reprinted with permission from Amgen Inc.

Abbreviations: BLA – Biologic License Application, FIPCO – fully independent pharmaceutical company, HCP – health care professional (e.g., physician, pharmacist, nurse), NDA – New Drug Application.

Support groups include important functional needs for operations of any company, such as, Finance, Human resources, Investor relations, Legal, Medical affairs, Public relations, and Regulatory, as identified in Table 10.3. Several key roles are specified for the elucidated departments in the table.

Product research and development in biopharma industry

The research and development process in the biopharma industry is a long range undertaking, often over 10+ years, with an extensive list of projects to be accomplished divided into 8 major steps involved: (1) Discovery research [over 2–5 years]; (2) Preclinical research [1–3 years]; (3) Submission of the Investigational New Drug application (IND) in the United States or Clinical Trials Application (CTA) in Europe [1/2–1 year]; (4) Review of Investigational application by regulatory agencies [1-1/2 years]; (5) Clinical Trials [3–5 years]; (6) Preparation of applications for licensing of products, for example, Biologics License Application/New Drug Application (in the United States)/Clinical Trials Document (in Europe)/Investigational Drug Exemption (BLA/NDA/CTD/IDE, respectively) [1 year]; (7) FDA Center for Drug Evaluation and Review (CDER) or Center for Biologics Evaluation and Research (CBER) review of marketing applications for

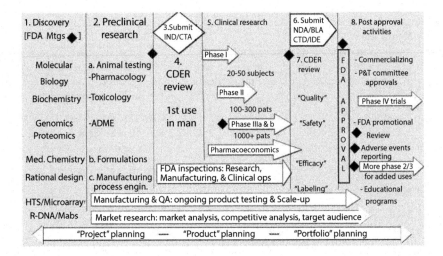

Figure 10.9 Drug development process - Extent and complexity.

Abbreviations: ADME – absorption, distribution, metabolism, and excretion studies, BLA – Biologics License Application in the United States, CDER – Center for Drug Evaluation and Review (in FDA), CTA – Clinical Trials Application (in Europe), CTD – Clinical Trials Document (in Europe), Engin – engineering, FDA – U.S. Food and Drug Administration, HTS – high-throughput screening, IDE – investigational drug exemption (for devices), IND – Investigational New Drug application (in the United States), Mabs – monoclonal antibodies, Med. – medical, mtgs – meetings, NDA – New Drug Application in the United States, ops – operations, pats – patients, P&T – Pharmacy and Therapeutics committee, QA – quality assurance. (Reprinted with permission from Amgen Inc.)

approval [1/2–1 year]; and (8) post-approval activities [3–10 years]. See Figure 10.9 for the summaries of selected projects within each of the 8 steps. Meetings are desired by biopharma companies with regulatory authorities, such as the FDA in the United States, at key times during R&D for feedback, guidance and approvals, which are signified by a diamond in Figure 10.9. Commonly biopharma companies have a scientific advisory board, which will meet 1–4 times a year to review the plans, clinical trials, and outcomes at multiple stages for a product and portfolio, as a means to obtain external validations and potential course corrections.

The main goal of the development process is a BLA or NDA, biologics license application or new drug application, respectively, which includes all the research data and reports (all phases) to establish sufficient efficacy, safety and quality of a product. Additionally, four initiatives occur over the life of a molecule from discovery through post marking, as represented in the figure by long arrows: (1) *FDA inspections and reviews* can and do occur at any and all steps, especially regarding manufacturing, clinical trials, applications, and marketing materials; (2) *manufacturing scale-up* from lab test materials, to preclinical study materials in animals (initial formulations of products created in preclinical stage), to ever-larger number of product doses for ever-larger phases of clinical trials, to full scale-up for marketing [plus quality testing at several steps ensuring product quality; (3) *market research* examines market potential of the product at multiple times as more data and more complete data are obtained as the product works its way through R&D. Other key areas of market research are target audiences, competitive product evaluations, and health care system evaluation; and (4) over the lifespan of a product, *planning and related project management* is performed for individual projects within a product and within departments, for products, and for the portfolio of products across all departments and divisions.

A "robust" pipeline of molecules that eventually become products in R&D is a necessity for company success. The ideal set of characteristics of "robustness" involve both the molecules and the process and number about 20 issues and needs, as follows for molecules: having (1&2) a full pipeline – molecules in both all stages of research and all stages of development, (3) a good number of molecules per stage, (4) robust data for efficacy (more and better), (5) robust data for toxicity (less in type and number), (6) novel molecules with unique targets or unique mechanisms of action, (7) chronic disease products (long-term use per patient = larger market potential), (8) diseases with large populations, (9) disease impact-disrupting pathogenesis or preventing mortality, (10) follow-on second and third indications pursued, (11) employing formulation improvements for patients and health care professionals, and (12) having strong patent positions. Process issues for robustness include (13) matching molecules to business units, (14) keeping work on schedule, (15) having necessary and appropriate resources (staff and grants) at the right times, (16) having systems in place for data processing, (17) using project and product management systems for efficiency and timeliness, (18) killing molecules early if questionable (resource optimization), (19) having an effective decision-making system (all the key issues addressed, all the necessary parties engaged, clear criteria, efficiently performed), and (20) operating and integrating globally.

The cost of operations at a biopharma company has been repeatedly studied and presents an ever-growing cost to research and develop for marketing a product to be and is estimated as of 2014 to exceed $2.5 billion per marketed product (including cost of capitalization). In the decade from 2003 to 2014, the growth in costs was 219%, from $802 million to $2.558 billion (see Figure 10.10). The substantial increase in R&D-related cost were larger trials required (more patients), which were much more complex trials with many more data points with more sophisticated testing, and higher per-patient fees to the physicians and health care institutions. The timeframe for all the work components for an application for product approval from research, product testing in laboratories, through

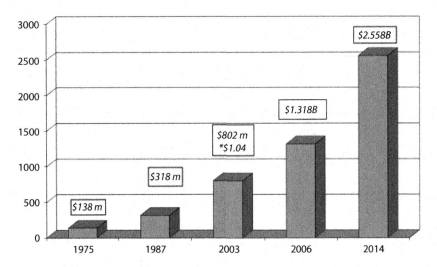

Figure 10.10 Estimated costs over time to develop and market products.

Abbreviations: B – billion in U.S. dollars, m – million in U.S. dollars. (Reprinted with permission from Tufts University.)

Phase 3 clinical trials, collectively often requires 8–10 years. The manufacturing for biotech products is a further expensive proposition. Two cost models are displayed in Figure 10.10 (Tufts University system) using data inputs from multiple companies and Figure 10.11 (Eli Lilly company system, actual company costs). Both cost models incorporate standard financial estimates and include actual costs to operate (out-of-pocket) and the cost of capital (Table 10.4).

Figure 10.11 Decision criteria.

Abbreviations: Metab – metabolism, P'kin – pharmacokinetics. (Reprinted with permission from Amgen Inc.)

Table 10.4 Products – times – costs for Eli Lilly modeling of R&D costs

Steps: Target · Hit · Lead · Lead Optim. · Candidate · Preclinical · Develop. (P.1-2-3) · BLA

Stage →	T-H	H-L	L.O.	Prelin. P/M/T	Ph.1	Ph.2	Ph.3	BLA Subm	Launch Totals
# Works In Progress	24	19	15	12	9	5	2	1	1 NME
Cycle Time	1 yr	1.5 yr	2 yr	1 yr	1.5 yr	2.5 yr	2.5 yr	1.5 yr	13 yrs
Cost OOP	$24m	49m	146m	62m	128m	185m	235m	44m	$873
Cost Capitalizatn	$94m	166m	414m	150m	273m	319m	314m	48m	$1,778
Total Costs	118m	215m	560m	212m	401m	504m	549m	92m	$2.651B

Source: Reprinted with permission from Nature Biotechnology.

Abbreviations: B – billions, BLA – Biologics License Application, Capitalizatn – capitalization, H – hit, L – lead, L.O. – lead optimization, m – millions, NME – new molecular entity, OOP – actual out-of-pocket expenses, P or Ph – phase of a clinical trials, P/M/T – pharmacodynamics, metabolism, toxicology, R&D – research and development, Subm – submission, T – target, yr – year.

BioPharma success factors – additional areas

Success in the biopharma industry (Figure 10.11) involves many key organizational and operational factors and science and business practices, a daunting set of challenges. *"Novel molecules"* from dramatic advances in science, along with *"operational excellence,"* are two cornerstones of biotechnology's success in bringing medical advances forward for improved patient care. Additionally, four factors were reviewed at length above that also comprise primary critical success factors, that is, the biotech's innovative *"culture," "financing"* growth, replication of tried-and-true *"structures of companies,"* efficient *"R&D operations."* Furthermore, five more success factors are suggested in following commentary. First, *"teams and decisions"* are two more important hallmarks for success for the biopharma industry. Especially in biotech, a team of individuals representing key multidisciplinary groups are empowered to plan, coordinate, lead, and decide the product's evolution, in a dynamic, iterative, interactive, demanding (of each other) modus operandi, through its complete life cycle from discovery through development and after marketing, that is, steps of Target-Hit-Lead-Candidate-Preclinical research, plus Manufacturing, Clinical research (Phases 1–3), Approvals, Sales, Follow-on indications). The members of the team usually include research, development, manufacturing,

marketing, regulatory affairs, legal, medical affairs, and finance, addressing any and all questions about a molecule related to the medical and
business world it exists within. Leadership, the team leader, often comes
from R&D early in a product's life and then S&M later in the life cycle.
Decisions of "Go – No Go" need to made at many key milestones in the
product's life cycle employing a standard and critical set of 10 questions/
parameters (Figure 10.11). Additionally, the teams consider these 10 decision criteria through four filters: (1) "What" is the "data" and its novelty;
(2) "Competition," other products on the market and any molecule in
development, from which other companies, regarding the United States
and worldwide; (3) for all projects, "When" will the work be done (dead-
line being met) and "by Whom" (internally or by vendors); and (4) "What"
has been "done" and "What else" needs to be done. The team makes pre-
liminary decisions to advance or terminate its single molecule/product of
focus, which are confirmed by senior management, considering also the
full portfolio of molecules, as well as individual products.

Second, *"planning and project management"* are two further critical success areas for biopharma industry. (1) Planning for any product or even an
overall portfolio must occur at multiple levels of the organization, (2) must
be well integrated among a plan's areas of consideration, (3) must engage
and integrate all divisions or departments as necessary, (4) must involve

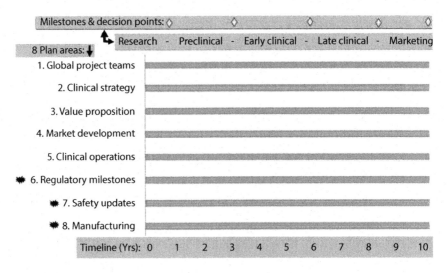

Figure 10.12 Planning at the corporate level. Asterisks indicate three plan areas
(regulatory, safety, and manufacturing) with the most impact that can create possible delays in the planning timeline. (Reprinted with permission from Springer
Publications.)

key milestones for advancement, (5) allow for adjusting plans over time, with (6) tracking in order to address timeliness. See Figure 10.12 for a sample planning document for any individual product, as well as the overall portfolio, identifying eight key areas, the milestones, and a 10-year timeline. Key decision points in a plan involve the milestones in advancing a molecule from its research to marketing and beyond (e.g., later added indications or dosage forms). The figure also suggests three plan areas with the most impact that can create possible delays in the planning timeline with an asterisk (regulatory, safety, and manufacturing). The highest level of planning is corporate portfolio, followed by individual products, individual departments, and individual projects.

Fourth, *project management* for a product is often quite detailed, addressing all the projects from all departments, employing chart systems such as a Gantt-style. Figure 10.13 presents 16 such project areas for one potential product, planned to be done over time (9 years), with dates for start times, duration of each project, deadlines identified by asterisks, termination times by circles and biannual checkpoints, represented by the vertical lines in the figure.

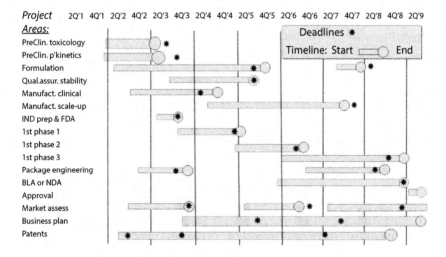

Figure 10.13 Project management chart. Dates for start times and duration of each project are indicated by boxes, deadlines are identified by asterisks and termination times by circles.

Abbreviations: BLA – Biologics License Application, FDA – U.S. Food and Drug Administration, IND – Investigational New Drug application, Manufact. – manufacture, NDA – New Drug Application, P'kinetics – pharmacokinetics, PreClin – preclinical, Qual.assur. – quality assurance. (Reprinted with permission from Springer.)

Overall measures of biotech success – product approvals, company acceptance, and sales performance

The biotechnology industry started in the 1970s with early company formation and now numbers over 4,000 companies worldwide, of which about 100 market a biotech product. Product approvals exceed 400, span an impressive breadth of types of molecules (proteins, peptides, monoclonal antibodies, vaccines, oligonucleotides, cell therapies, tissue therapies, gene therapy), and include about 370 different indications engaging all medical disciplines. Figure 10.14 displays all FDA product approvals from 1990 to 2018 in 5-year segments for novel new drugs (NDAs) and novel biotech products (BLAs and NDAs), along with biotech products as a percentage of total products. From the years 2000–2004 to the present, biotech products account for 30%–40% of all approvals.

Also, in the last 6 years (2014–2019), success has been dramatic in biotech with product approvals exceeding 20 per year (average of 31 per year with a range of 22–37) and 19–58 new indications (average of 24) per year. The novelty of drugs and biotech products was studied by FDA staff (Figure 10.15), covering 903 product/drug approvals. They demonstrated

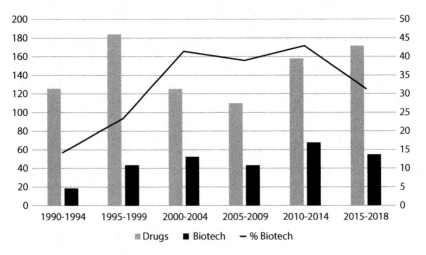

Figure 10.14 Novel FDA NDAs and BLAs with Biotech percentage of all approvals. Biotech approvals in this tabulation exclude biosimilars, vaccines, peptides that were second or later product approvals, and tissue therapies that were excluded by the FDA because they were not considered novel or were not included in their biotech tabulations.

Abbreviations: BLA – Biologics License Application, FDA – U.S. Food and Drug Administration, NDA – New Drug Application. (Reprinted and updated with permission from Tufts University.)

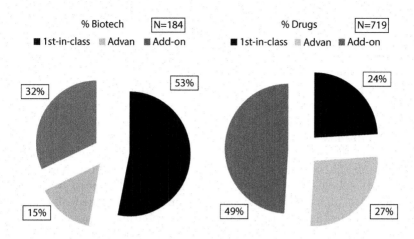

Figure 10.15 Innovation of NMEs and NBEs.

Abbreviations: Advan – advantage, NBE – new biological entity, NME – new molecular
entity. (Reprinted with permission from ASCPT.)

that biotech products were characterized as substantially more innovative
in their health care impact, comparing three categories: "first-in-class,"
versus "advantage" for the new product, versus only an "add-on" product
with no distinct advantage. Biotech products were innovative for 53% of
the products versus 24% of the time for drugs.

A successful biotech company will be engaged in *"alliances and mergers and acquisitions" (M&As)* with another often larger, established company who recognizes the medical advances and the significant marketing
opportunity of the biotech's technology and products. In alliances, the
primary agreement is the biotech company out-licenses a product, or a
group of similar products, or a technology (novel assets) to a larger company for their use to develop and market the products going forward.
In return, the biotech company receives cash as revenue to continue their
operations and also possibly receives a percentage of future sales or a roy-
alty. The biotech company may also receive important access to the expertise of the larger established company regarding development, regulatory,
and marketing areas of expertise, where their own experiences are very
limited. The reputation of the biotech company is enhanced with an alliance, through an external validation of the value and utility of its science
and product market opportunities. Further, the R&D and BLA could be
accelerated with the added resources devoted to the product out-licensed
to the established large, experienced company. The licensee (pharma company) obtains in an alliance the novel asset along with their patents, new
markets, and new potential profits. Plus, the licensee's motivations may
be due to expiring patents on their existing products, a relatively weak

internal pipeline needing invigorating, and a stash of cash available for transactions. Alliances account annually for many collaborations and are the largest source of revenue for biotech companies prior to sales of products. Over the recent years of 2012 through 2018, one tabulation of alliances for licensing-type deals showed 88–150 alliances in any one year at cost of total cost of $3.03–$8.61 billion per year. Also, the largest revenue source for public biotech companies is partnerships with other companies, usually 40%–50% of total revenues.

Instead of only acquiring access to specific products or technologies, a large pharma company or large biotech may decide to purchase, that is, acquire the whole company. M&As offer them the full biotech pipeline and technologies, plus access to the brain-power, that is, the scientists performing the work, which can further enhance the pharma company's capabilities in new scientific areas. M&As are quite common, such that over the

Table 10.5 Worldwide mergers & acquisitions of biotech by pharma 1998–2019

Year	No. deals	No. Bb	$B	Mega deals ($5B+)
2018–2019	35	18	220	Sanofi-Bioverativ, Novartis-AveXis & -THC (The Medicines Company), Takeda-Shire, Celgene-Juno, BMS-Celgene
2013–2017	79	25	290	Perrigo-Elan, Merck-Cubist, Roche-Intermune, Actavis-Allergan, Shire-NPS, AbbVie-Pharmacyclics, Pfizer-Medivation, Shire-Baxalta, Lilly-Stemcentx, Gilead-Kite, Gilead-Pharmasset, Takeda-Ariad
2012–2008	83	23	226	BristolMyersSquibb-Amylin, Sanofi-Genzyme, Teva-Cephalon, Teva-Ratiopharm, Roche-Genentech (49%), Pfizer-Wyeth, Takeda-Millenium, Lilly-lmclone
2007–2003	46	19	81	AstraZeneca-MedImmune, Merck-Organon, MerckKGaA-Serono
2002–1998	18	5	27	Johnson & Johnson-Alza, Roche-Genentech (51%)

Source: Reprinted and updated with permission from CPT.

Note: For the pairs of companies listed in the M&A deals, the first company listed is the acquirer and the second name is the company being acquired.

Abbreviations: B – billion U.S. dollars, Bb – blockbuster deals (>$1 billion), BristolMyersSquibb-Amylin – Bristol-Myers-Squibb-Amylin.

last 22 years (1998 to 2019), 41 pharma companies have acquired 261 biotech companies costing a total of $844 billion, especially in last 12 years, as represented in Table 10.5. Blockbuster deals numbered 244 (at least $1 billion price tag). The mega M&A deals of at or over $5 billion are enumerated below (#30).

Biotech product *"usage and sales"* has grown dramatically and consistently over time from the early days to the present timeframe, and such growth represents an overall measure of exceptional success of the biotech industry (Figure 10.16). In 2002, biotech sales reached $36 billion, 3.7% of all drug prescription sales, and by 2017 sales achieved about $241 billion, 22% of all drug prescription sales. Another indicator of this dramatic impact of biotechnology products on health care and all drug sales is the list of the top 25 blockbuster products (at least $1 billion annual sales) in 2018, wherein drugs number only five of them (noted in table with italics print) versus 20 for biotech (20% vs. 80%) and total sales of $35.2 billion (22.4%) vs. $121.7 billion (77.6%), respectively (see Figure 10.16). The oncology area leads product sales by far in number of products (#8) and sales dollars 35.2%, $55.2 billion of $156.9 billion). Immune disease products used in dermatology, gastroenterology, and rheumatology account for five products and $42.9 billion (27.3%) (Table 10.6).

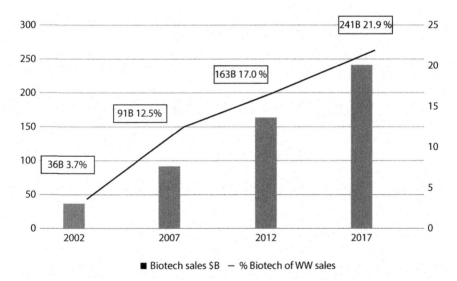

Figure 10.16 Biotech product sales over time, 2002 to 2017. B – billon in U.S. dollars. *Abbreviation:* WW – worldwide. (Reprinted with permission from Tufts University.)

Table 10.6 Top 25 blockbuster product sales ($B) in 2018

Name	Med. area	Sales	Name	Med. Area	Sales
1. Humira	D/GE/Rh	20.359	14. Stelara	D/Rh	5.293
2. *Eliquis*[a]	CV	9.872	15. *Lyrica*	*Neuro*	4.970
3. *Revlimid*	*Oncol*	9.685	16. Neulasta	Oncol	4.493
4. Rituxan	Oncol	7.965	17. Lantus	Endo	4.214
5. Opdivo	Oncol	7.551	18. *Imbruvica*	*Oncol*	*4.100*
6. Enbrel	D/GE/Rh	7.443	19. Victoza	Endo	3.869
7. Keytruda	Oncol	7.171	20. Lucentis	Oph	3.745
8. Herceptin	Oncol	7.149	21. Botox	D/N	3.577
9. Avastin	Oncol	7.043	22. Gardasil	ID	3.500
10. Eylea	Oph	6.746	23. Simponi	GE/Rh	3.332
11. *Xarelto*	CV	*6.589*	24. Humalog	Endo	2.997
12. Remicade	D/GE/Rh	6.438	25. Xolair	Pulm	2.997
13. Prevna	ID	5.802			

Abbreviations: B – billion, D – Dermatology, CV – Cardiovascular, Endo – Endocrinology, GE – Gastroenterology, ID – Infectious disease, Med – Medical discipline, N or Neuro – Neurology, Oncol – Oncology, Oph – Ophthalmology, Pulm – Pulmonary, Rh – Rheumatology.

[a] The drugs in italics were among the top 25 blockbuster products.

Biotechnology products and utilization

Almost all biotechnology products are limited in their handling and use by their relative instability and susceptibility to degradation of the molecules in contrast to most drugs.

Most products are injectable and do not contain preservatives because often this ingredient can interact with and cause degradation of proteins, necessitating single-use vials. Most products require refrigeration to extend the stability and shelf life. Some products are formulated and produced in a lyophilized form for improved stability reasons as well. Temperature changes in the extremes of freezing and high heat cause more changes with biologicals versus drugs, again limiting handling. Proteins can be degraded or result in aggregation or clumping with excess agitation, further limiting handling. Diluents used in their preparation for administration can lead to instability, for example, two protein growth factors closely related in structure and use possess, however, different profiles in their administration, that is, filgrastim requires dextrose in water and not normal saline, whereas sargramostim can be used with normal saline. Close adherence to the usage guidelines in an official package circular for biotechnology products is a particular very important need. Finally, the cost of biotechnology products – which is quite high to the health care system relative to most drugs ($25,000–$50,000 per year for arthritis therapy, up to $100,000–200,000 for a course of treatment in some cancers, $400,000/ year or more for enzyme gene defect diseases, over $1 million for curative gene therapies) – must be weighed against the novelty of the molecules in their actions on disease pathogenesis and their substantial therapeutic advances.

Summary

The advancement and success of biotechnology from its fledgling days of product approvals in early 1980s to more current times in 2019 are measured well by the product approvals over time. Over 440 products and 375 indications have been approved by the FDA for biotech products. Over time the approval of new biotech products and indications has risen quite dramatically, as observed in Figure 10.17 displaying 5-year windows in time. In the last 5-year timeframe (2015 to 2019), approvals have exploded, almost doubling the prior 5 years.

Another measure of the success of biotech and especially its staying power is its product pipeline, represented by products in clinical development over time. The full engagement of the pharma and biotech companies in biotechnology is well represented in the molecules in clinical trials by each company. Table 10.7 presents 29 companies with at least 10 products up to 55 each in their respective biotech pipelines; a total of over 750 biotech products in clinical trials by only these 29 companies. Biotech companies (11) are noted with bold print. Furthermore, Figure 10.18

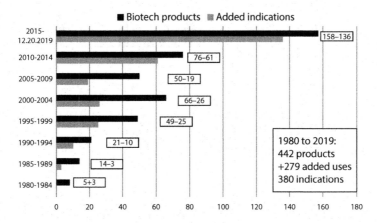

Figure 10.17 Biotech product approvals over time (1980 to 2019). (Reprinted with permission from Tufts University.)

Table 10.7 Companies with biotech products in clinical trials in 2018

Company	No. trials	Company	No. trials	Company	No. trials
1. Sanofi	55	11. JnJ	31	21. Boehr.-Ing.	14
2. Novartis	51	12. **Ionis**	30	22. Bayer	13
3. GSK	50	13. Merck	28	23. **Celldex**	13
4. Pfizer	48	14. AbbVie	27	24. **BioMarin**	12
5. Roche[a]	44	15. Celgene	27	25. Merck KGaA	12
6. BMS	44	16. **Regeneron**	21	27. Allergan	11
7. Eli Lilly	43	17. Astellas	20	28. **Inovio**	11
8. Astra-Zeneca	40	18. **Novo Nord.**	19	29. **Xencor**	10
9. **Amgen**	40	19. **Biogen**	18		
10. **Shire**	37	20. **Seattle Gen.**	16		

Abbreviations: Boehr. Ing.– Boehringer Ingelheim, BMS – Bristol-Myers Squibb, GSK – GlaxoSmithKline, JnJ – Johnson & Johnson, Novo Nord. – Novo Nordisk, Seattle Gen. – Seattle Genetics.

[a] Biotech companies are noted in bold print.

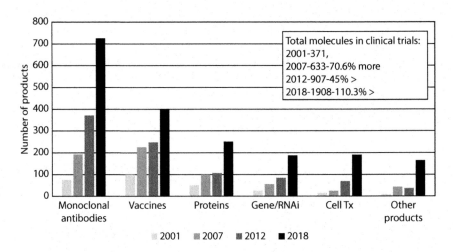

Figure 10.18 Biotech products in clinical trials (Phase 1, 2, & 3) (as of December 31, 2018).

Abbreviations: RNAi – RNA inhibition, Tx – treatment. (Reprinted with permission from Tufts University.)

displays both the impressive broad scope of products and their significant growth over time in the number of products, as well as in each category. Monoclonal antibodies are the lead product molecule type, with an over-all growth from 371 molecules in 2001 to 1908 in 2018.

Bibliography

Chapter 1

Anonymous. *Timeline (Biotechnology)*. Biotechnology Industry Organization, Washington, DC, 2006. www.bio.org/speeches/pubs/er (accessed March 1, 2006).

Anonymous. *Timeline of Medical Biotechnology*. Amgen Inc., Thousand Oaks, CA, 2019.

Bates SL, Zhao J-Z, Roush RT, Shelton AM. Insect resistance management in GM crops: Past, present, and future. *Nat Biotechnol* 2005;23:57–62.

Colwell B. Biotechnology timeline: Humans have manipulated genetics since dawn of civilization. *Genetic Literacy Project.org*, July 18, 2017.

Evens RP. *Biotechnology and Biological Products, Encyclopedia of Pharmaceutical Technology*. 2nd ed., Swarbrick J, Ed., Marcel Dekker, New York, 2001.

Evens RP. *Biotechnology and Biological Products, Encyclopedia of Pharmaceutical Technology*. 3rd ed., Swarbrick J, Ed., Marcel Dekker InForma Healthcare, New York, pp. 258–280, 2007.

Evens RP. *Biotechnology and Biological Products, Encyclopedia of Pharmaceutical Technology*. 4th ed., Swarbrick J, Ed., Taylor & Francis Group, Boca Raton, FL, vol. 1, pp. 239–266, 2013.

Evens RP. *Comprehensive Biotechnology Data Book*. Tufts University, Center for the Study of Drug Development, Boston, MA, 2020 (in press).

Evens RP, Dinarello CA, Browne J, Fenton D. Biotechnology and clinical medicine: Part I. *Hosp Physician* 1995;31:27–36.

Evens RP, Dinarello CA, Browne J, Fenton D. Biotechnology and clinical medicine: Part II. *Hosp Physician* 1995;31:26–31.

Evens RP, Kaitin KI. Evolution of the biotechnology and its impact on healthcare. *Health Affairs* 2015;34(2):210–219.

Genetics and genomics timeline. *Genome News Network*. Genomenewsnetwork.org/resources/timeline/1973.

Giovannetti G, Ed. *Beyond Borders: Reaching New Heights. Biotechnology Industry Report 2015*. Ernst & Young, San Francisco, CA.

Giovannetti GT, Jaggi G, Bialojan S, Eds. *Beyond Borders: Global Biotechnology Report*. Ernst & Young, San Francisco, CA, 2010.

Herdt RW. Biotechnology in agriculture. *Ann Rev Environ Resour* 2006;31(11): 265–295.

Herrera S. Industrial biotechnology – a chance at redemption. *Nat Biotechnol* 2004;22:671–675.

Huggett B, Hodgson J, Lahteenmaki R. Public biotech 2010 – The numbers. *Nat Biotechnol* 2011;29:585–591.

ISAAA. Pocket K No. 16: Biotech crops. Highlights in 2017. isaaa.org/resources/publications/pocket/16.

Kashanguru C. Industrial biotechnology: Then, now and the future. *Conference: 2nd Training Workshop on Industrial Biotechnology*, September 2018, Harare, Zimbabwe.

Kleter GA, van der Krieken WM, Kok EJ, et al. Regulation and exploitation of genetically modified crops. *Nat Biotechnol* 2001;19:1105–1110.

Lahteenmaki R, Fletcher L. Public biotech 2000 – The numbers. *Nat Biotechnol* 2001;19:407–411.

Lahteenmaki R, Lawrence S. Public biotech 2005 – The numbers. *Nat Biotechnol* 2006;24:625–634.

Lawrence S. Agbio keeps on growing. *Nat Biotechnol* 2005;23:281.

Mannion AM, Morse S. Biotechnology in agriculture: Agronomic and environmental considerations and reflections based on 15 years of GM crops. *Prog Phys Geogr* 2012;36(6):747–763.

Marshall A. 13.3 million farmers cultivate GM crops. *Nat Biotechnol* 2009;27:221.

Morrison C, Lahteenmaki R. Public biotech 2015 – The numbers. *Nat Biotechnol* 2016;34:709–715.

Norero D. GMO crops have been increasing yields for 20 years, with more progress ahead. *Cornell Alliance for Science*, February 23, 2018.

Peng W. GM crop cultivation surges, but novel traits languish. *Nat Biotechnol* 2011;29(4):302.

Szaro D, Ed. *Beyond Borders: A Global Perspective*. Ernst & Young, San Francisco, CA, 2005.

Szaro D, Ed. *Beyond Borders: Global Biotechnology Report 2000*. Ernst and Young, San Francisco, CA, 2000.

Walsh G. Biopharmaceutical benchmarks 2006. *Nat Biotechnol* 2006;24:769–736.

Walsh G. Biopharmaceutical benchmarks 2010. *Nat Biotechnol* 2010;28:917–924.

Walsh G. Biopharmaceutical benchmarks 2014. *Nat Biotechnol* 2014;32:992–1000.

Walsh G. Biopharmaceutical benchmarks 2019. *Nat Biotechnol* 2018;37:1136–1145.

Watanabe ME. Can bioremediation bounce back? *Nat Biotechnol* 2001;19:1111–1115.

Wieczorek AM, Wright MG. History of agricultural biotechnology: How crop development has evolved. *Nat Educ Knowl* 2012;3(10):9.

Chapter 2

Anonymous. *Biotechnology Tools in Research and Development*. Biotechnology Industry Organization, Washington, DC, 2006a. www.bio.org/speeches/pubs/er (accessed March 1, 2006).

Anonymous. *The Technologies and Their Applications (Biotechnology)*. Biotechnology Industry Organization, Washington, DC, 2006b. www.bio.org/speeches/pubs/er (accessed March 1, 2006).

Carroll WL. Introduction to recombinant-DNA technology. *Am J Clin Nutr* 1993;58:249S–258S.

Demain AL. An overview of biotechnology. *Occup Med* 1991;6:157–168.

Dutton G. Biologics force changes in stability testing. *Gen Eng News* 2007;27:47–49.

Evens RP. *Biotechnology and Biological Products, Encyclopedia of Pharmaceutical Technology.* 4th ed., Swarbrick J, Ed., Taylor & Francis Group, Boca Raton, FL, vol. 1, pp. 239–266, 2013.

Evens RP, Dinarello CA, Browne J, Fenton D. Biotechnology and clinical medicine: Part I. *Hosp Physician* 1995;31:27–36.

Evens RP, Kaitin KI. The evolution of biotechnology and its impact on health care. *Health Affairs* 2015;34(2):210–219.

Evens RP, Witcher M. Biotechnology: An introduction to recombinant DNA technology and product availability. *Ther Drug Monit* 1993;15:514–520.

Flanagan N. Protein therapeutic formulation issues. *Gen Eng News* 2010;30:32–37.

Frokjaer S, Otsen DE. Protein drug stability: A formulation challenge. *Nat Rev Drug Discov* 2005;4:298–306.

Khan S, Ullah MW, Siddque R, et al. Role of recombinant DNA technology to improve life. *Int J Genomics* 2016;2016:1–14.

Kiss RD. *Gen Eng News* 2004;24(8):1.

Leader B, Baca QJ, Golan DE. Protein therapeutics: A summary and pharmacological classification. *Nat Rev Drug Discov* 2008;7:21–39.

Molowa DT, Mazanet R. The state of biopharmaceutical manufacturing. *Biotechnol Annu Rev* 2003;9:285–302.

Pallanathu MK. *Gen Eng News* 2006;26(1):552.

Schneider CK, Schaffner-Dallman G. Typical pitfalls in applications for marketing authorization of biotechnology products in Europe. *Nat Rev Drug Discov* 2008;7:893–899.

Chapter 3

Anonymous. The evolution of antibody-drug manufacturing. SAFE. Millipore Sigma, Merck KGaA, September 2018.

Baker M. Upping the ante on antibodies. *Nat Biotechnol* 2005;23:1065–1072.

Carson KI. Flexibility – The guiding principle for antibody manufacturing. *Nat Biotechnol* 2005;23(9);1054–1058.

Dhara VG, Naik HM, Majewska NI, Betenbaugh MJ. Recombinant antibody production in CHO and NS0 cells: Differences and similarities. *BioDrugs* 2018;32(6):571–584.

Garber K. No-added sugar: Antibody makers find an upside to no fucose. *Nat Biotechnol* 2018;36(11):1025–1027.

Holliger P, Hudson PJ. Engineering antibody fragments and rise of single domains. *Nat Biotechnol* 2005;23:1126–1136.

Hoogenboom HR. Selecting and screening recombinant antibody libraries. *Nat Biotechnol* 2005;23:1105–1116.

Hughes B. Antibody-drug conjugates for cancer: Poised to deliver? *Nat Rev Drug Discov* 2010;9:665–669.

Jiang X-R, Song A, Bergelson S, et al. Advances in the assessment and control of the effector functions of therapeutic antibodies. *Nat Rev Drug Discov* 2011;10:101–110.

Labrijn AF, Janmaat ML, Reichert JM, Parren PWHI. Bispecific antibodies: A mechanistic review of the pipeline. *Nat Rev Drug Discov* 2019;18(8):585–608.

Leavy O. Therapeutic antibodies: Past, present and future. *Nat Rev Immunol* 2010;10(5):297–305.

Morrison C. Nanobody approval gives domain antibodies a boost. *Nat Rev Drug Discov* 2019;18(7):485–487.

Mould DR, Green B. Pharmacokinetics and pharmacodynamics of monoclonal antibodies. *BioDrugs* 2010;24:23–39.

Muller D, Kontermann RE. Bispecific antibodies for cancer immunotherapy. *BioDrugs* 2010;24:89–98.

Nelson AL, Dhimolea E, Reichert JM. Development trends for human monoclonal antibody therapeutics. *Nat Rev Drug Discov* 2010;9:767–774.

Nelson AL, Reichert JM. Development trends for therapeutic antibody fragments. *Nat Biotechnol* 2009;27:331–337.

Reichert JM, Valge-Archer VE. Development trends for monocloncal antibody cancer therapeutics. *Nat Rev Drug Discov* 2007;6:349–356.

Saldeld JG. Isotype selection in antibody engineering. *Nat Biotechnol* 2007;25:1369–1372.

Sauerborn M, van Dongen W. Practical considerations for the pharmacologic and immunogenic assessment of antibody-drug conjugates. *BioDrugs* 2014; 28:383–391.

Shankar G, Arkin S, Devanarayan V, et al. The quintessence of immunogenicity reporting for biotherapeutics. *Nat Biotechnol* 2015;33(4):334–335.

Singh S, Kumar NK, Dwiwedi P, et al. Monoclonal antibodies review. *Curr Clin Pharmacol* 2018;13(2):85–99.

Wang Y-M, Wang J, Hon YY, Zhou L, Ahn HY. Evaluating and reporting the immunogenicity impacts for biological products – A clinical pharmacology perspective. *AAPS J* 2016;18(2):395–403.

Wu AM, Senter PD. Arming antibodies: Prospects and challenges for immuno-conjugates. *Nat Biotechnol* 2005;23:1137–1146.

Chapter 4

Bulbake U, Doppalapudi S, Kommunineni N, Khan W. Liposomal formulations in clinical use: An updated review. *Pharmaceutics* 2017;9(12):1–33.

Davis ME, Chen Z, Shin DM. Nanoparticle therapeutics: An emerging treatment modality for cancer. *Nat Rev Drug Discov* 2008;7:771–782.

Dimond PF. Using nanotechnologies in biotech and medicine. *Gen Eng News* 2005;25:1–21.

Evens RP. Vaccine products in R&D pipeline have tripled since 2005. *Tufts CSDD Impact Report* 2015;17(4):1–4.

Fosgerau K, Hoffman T. Peptide therapeutics: Current status and future directions. *Drug Discov Today* 2015;20(1):122–128.

Garber K. Peptide leads new class of chronic pain drugs. *Nat Biotechnol* 2005;23:399.

Gilboa E. The promise of cancer vaccines. *Nat Rev Cancer* 2004;4(5):401–411.

Glaser V. Vaccine production shifts to high gear. *Gen Eng News* 2009;29:1,52, 53–56.

Goetz KB. Pfleiderer M, Schneider CK. First-in-human clinical trials with vaccines – What regulators want. *Nat Biotechnol* 2010;9:910–916.

Goldman B, DeFrancesco L. The cancer vaccine roller coaster. *Nat Biotechnol* 2009;27:129–139.

Kresse H, Rovini H. Influenza vaccine market dynamics. *Nat Rev Drug Discov* 2009;8:841–842.

Kresse H, Shah M. Strategic trends in the vaccine market. *Nat Rev Drug Discov* 2010;9:913–914.

Lau LL, Dunn MK. Therapeutic peptides: Historical perspectives, current development trends and future directions. *BioOrg Med Chem* 2018;26:2700–2707.

Lutolf MP, Hubbell JA. Synthetic biomaterials as instructive microenvironments for morphogenesis in tissues engineering. *Nat Biotechnol* 2005;23:47–55.

Mazzola L. Commercializing nanotechnology. *Nat Biotechnol* 2003;21:1137–1143.

McNeish J. Embryonic stem cells in drug discovery. *Nat Rev Drug Discov* 2004;3:70–80.

Paull R, Wolfe J, Hebert P, Sinkula M. Investing in nanotechnology. *Nat Biotechnol* 2003;21:1144–1147.

Petit-Zeman S. Regenerative medicine. *Nat Biotechnol* 2001;19:201–206.

Porter CJH, Trevaskis NL, Charman WN. Lipids and lipid-based formulations: Optimizing the oral delivery of lipophilic drugs. *Nat Rev Drug Discov* 2007;6:231–248.

Preti RA. Bringing safe and effective cell therapies to the bedside. *Nat Biotechnol* 2005;23:801–804.

Reichert JM. Therapeutic peptides in clinical study in 2000–07 nearly doubled 1990s rate. *Tufts Center for the Study of Drug Development. Impact Report* 2009;11:1–4.

Sheridan C. Flu vaccine makers upgrade technology – and pray for time. *Nat Biotechnol* 2009;27:489–491.

Sheridan C. Proof of concept for next-generation nanoparticle drugs in humans. *Nat Biotechnol* 2012;30(6):471–473.

Sheridan C. Vaccine market boosters. *Nat Biotechnol* 2009;27:499–501.

Shmulewitz A, Langer R, Patton J. Convergence in biomedical technology. *Nat Biotechnol* 2006;24:277–281.

Taroncher-Oldenburg G. Vaccine dealmaking set for further growth in 2015. *Nat Rev Drug Discov* 2015;14(3):B19–B21.

Torchilin V. Recent advances with liposomes as pharmaceutical carriers. *Nat Rev Drug Discov* 2005;4:145–160.

Torchilin VP. Multifunctional, stimuli-sensitive nanoarticle systems for drug delivery. *Nat Rev Drug Discov* 2014;13(11):813–827.

Whitesides GM. The 'right' size in nanobiotechnology. *Nat Biotechnol* 2003;21:1161–1165.

Chapter 5

Anselmo AC, Gokarn Y, Mitragotri S. Non-invasive delivery strategies for biologics. *Nat Rev Drug Discov* 2019;18(1);19–40.

Araujo RP, Liotta LA, Petricoin EF. Proteins, drug targets and the mechanisms they control: The simple truth about complex networks. *Nat Rev Drug Discov* 2007;6:871–880.

Baldo BA. Chimeric fusion proteins used for therapy: Indications, mechanisms and safety. *Drug Saf* 2015;38(5):455–479. doi:10.1007/s40264-015-0285-9.

Belsey MJ, Harris LM, Das RR, Chertkow J. Biosimilars: Initial excitement gives way to reality. *Nat Rev Drug Discov* 2007;5:535–536.

Brooks SA. Appropriate glycosylation of recombinant proteins for human use: Implications of choice of expression system. *Mol Biotechnol* 2004;28:241–255.

Czajkowsky DM, Hu J, Shao Z, Pleass RJ. Fc-Fusion proteins: New developments and future perspectives. *Mol Med* 2012;4(10):1015–1028. doi:10.1002/emmm.201201379.

Dove A. The bittersweet promise of glycobiology. *Nat Biotechnol* 2001;19:913–917.

Ekladious I, Colson YL, Grinstaff MW. Polymer-drug conjugate therapeutics: Advances, insights and prospects. *Nat Rev Drug Discov* 2019;18(4):273–294.

Furlani P. Biosimilars: A regulatory review (EU). *Drug Inform J* 2008;42:477–485.

Geysen HM, Schoenen F, Wagner D, Wagner R. Combinatorial compound libraries: An ongoing challenge. *Nat Rev Drug Discov* 2003;2:222–230.

Gottlieb S. Biosimilars: Policy, clinical, and regulatory considerations. *Am J Health Syst Pharm* 2008;65(Suppl 6):52–58.

Grabowski H. Follow-on biologics: Data exclusivity and the balance between innovation and competition. *Nat Rev Drug Discov* 2008;7:479–488.

Graddis TJ, Remmele RL, McGrew JT. Designing proteins that work using recombinant technologies. *Curr Pharm Biotechnol* 2002;3:285–297.

Harris JM, Martin NE, Modi M. Pegylation: A novel process for modifying pharmacokinetics. *Clin Pharmacokinet* 2001;40:539–551.

Hodgson J. WHO guidelines presage US biosimilars legislation? *Nat Biotechnol* 2009;27:963–965.

Jeffries R. Glycosylation as a strategy to improve antibody-based therapeutics. *Nat Rev Drug Discov* 2009;8:226–234.

Kingsmore SF. Multiplexed protein measurement: Technologies and applications of protein and antibody arrays. *Nat Rev Drug Discov* 2006;5:310–320.

Lagasse HAD, Alesaki A, Simhadri VL, et al. Recent advances (therapeutic protein) drug development. *F1000Res* 2017;6(2):113. doi:10.12688/f1000research.9970.1.

Lanthier M, Behrman R, Nardinelli C. Economic issues with follow-on protein products. *Nat Rev Drug Discov* 2008;7:733–737.

Macarron R, Banks MN, Bojanic D, et al. Impact of high throughput screening in biomedical research. *Nat Rev Drug Discov* 2011;10:188–195.

Miller KL, Lanthier M. Innovation in biologic new molecular entities: 1986–2014. *Nat Rev Drug Discov* 2015;14(2):83.

Molineaux G. Pegylation: Engineering improved biopharmaceutics for oncology. *Pharmacotherapy* 2003;23(8 Pt 2):3s–8s.

Morrow KJ. Developing sophisticated protein-based drugs engineering and expressing novel therapeutic proteins. *Gen Eng News* 2005;25:1,11–12.

Ratner M. Pharma swept up in biogenerics gold rush. *Nat Biotechnol* 2009;27:299–301.

Sanchez-Martin RM, Mittoo S, Bradley M. The impact of combinatorial methodologies on medicinal chemistry. *Curr Top Med Chem* 2004;4:653–669.

Schellekens H, Moors E. Clinical comparability and European biosimilar regulations. *Nat Biotechnol* 2010;1:28–31.

Schneider CK, Kalinke U. Toward biosimilar monoclonal antibodies. *Nat Biotechnol* 2008;26:985–990.

Sheridan C. Llama-inspired antibody fragments approved for rare blood disorder. *Nat Biotechnol* 2019;37(4):333–334.

Sola RJ, Griebenow K. Glycosylation of therapeutic proteins. *BioDrugs* 2010;24(1):9–21.

Spurr N, Darvasi A, Terrett J, Jazwinska L. New technologies and DNA resources for high throughput biology. *Br Med Bull* 1999;55:309–324.

Strohl WR. Fusion proteins for half-life extension of biologics as a strategy to make biobetters. *BioDrugs* 2015;29:215–239.

Veronese FM, Pasut G. Pegylation, successful approach to drug delivery. *Drug Discov Today* 2005;10:1451–1458.

Wenzel RG. Current legal, regulatory, and scientific implications of biosimilars. *Am J Health Syst Pharm* 2008;65(Suppl 6):S1–S22.

Woodcock J, Griffin J, Behrman R, et al. The FDA's assessment of follow-on protein products: A historical perspective. *Nat Rev Drug Discov* 2007;6:437–442.

Chapter 6

Adli M. The CRISPER tool kit for genome editing and beyond. *Nat Commun* 2018;9:1911–1923.

Bagheri S, Kashani-Sabat M. Ribozymes in the age of molecular therapeutics. *Curr Mol Med* 2004;4:489–506.

Baker M. In biomarkers we trust. *Nat Biotechnol* 2005;23:297–298.

Bernard S. The 5 myths of pharmacogenomics. *Pharm Exec* 2003;23:70–78.

Branca MA. Gene therapy: Cursed or inching towards credibility. *Nat Biotechnol* 2005;23:519–521.

Byrnes AP. Challenges and future prospects in gene therapy. *Drugs* 2005;8: 943–946.

Davis LC, Furstenthal L, Desai AA, et al. The microeconomics of personalized medicine: Today's challenge and tomorrow's promise. *Nat Rev Drug Discov* 2009;8:279–286.

de Fougerolles A, Vornlocher H-P, Maraganore J, Lieberman J. Interfering with disease: A progress report on siRNA-based therapeutics. *Nat Rev Drug Discov* 2007;6:443–453.

Dolgin E. 'Bubble' boy gene therapy reignites commercial interest. *Nat Biotechnol* 2019;37(7):699–701.

Dorsett Y, Tuschl T. siRNAs: Applications in functional genomics and potential genomics as therapeutics. *Nat Rev Drug Discov* 2004;3(4):318–329.

Dove A. Antisense and sensibility. *Nat Biotechnol* 2002;20:121–124.

El-Aneed A. Current strategies in cancer gene therapy. *Eur J Pharmacol* 2004;498:1–8.

Fitzsimmons LD. Biomarkers: Expanding their discovery. Hurry up and wait. *R&D Directions* 2007;13:16–23.

Foss GS, Rogne S. Gene medication or genetic modification? The devils in the details. *Nat Biotechnol* 2003;21:1280–1281.

Freelove AL, Zhang R. The power of ribozyme technologies: The logical way ahead for molecular medicine and gene therapy? *Curr Opin Mol Ther* 2002;4:419–422.

Garzon R, Marcucci G, Croce CM. Targeting microRNAs in cancer: Rationale, strategies and challenges. *Nat Rev Drug Discov* 2010;9:775–789.

Gordon K, Del Medico A, Sander I, et al. Gene therapies in ophthalmic disease. *Nat Rev Drug Discov* 2019;18(6):415–416.

High KA, Roncarolo MG. Gene therapy. *NEJM* 2019;381:455–464.

Higuchi Y, Kawakami S, Hashid M. Strategies for the in vivo delivery of siRNAs. *BioDrugs* 2010;24:196–205.

Hopkins MM, Ibarreta D, Caisser S, et al. Putting pharmacogenetics into practice. *Nat Biotechnol* 2006;24:403–410.

Keefe AD, Pai S, Ellington A. Aptamers as therapeutics. *Nat Rev Drug Discov* 2010;9:537–550.

Keeler AM, ElMallah MK, Flotte TR. Gene therapy 2017: Progress and future directions. *Clin Transl Sci* 2017;10(4):242–248.

Khan SH. Genome-editing technologies: Concept, pros, and cons of various genome-editing techniques and bioethical concerns for clinical application. *Mol Ther Nucleic Acids* 2019;16(6):326–334.

Kirchheiner J, Fuhr U, Brockmoller J. Pharmacogenetics-based therapeutic recommendations – Ready for clinical practice? *Nat Rev Drug Discov* 2005;4:639–647.

Kling J. First US approval for a transgenic animal drug. *Nat Biotechnol* 2009;27: 302–304.

Kole R, Krainer AR, Altman S. RNA therapeutics: Beyond RNA interference and antisense oligonucleotides. *Nat Rev Drug Discov* 2012;11(2):125–140.

Krejsa C, Rogge Sadee W. Protein therapeutics: New applications for pharmacogenetics. *Nat Rev Drug Discov* 2006;5:507–521.

LeMieux J. Going viral: The next generation of AAV vectors. *Gen Eng Biotechnol News* 2019;39(9):22–26.

Lesko LJ, Woodcock J. Translation of pharmacogenomics and pharmacogenetics: A regulatory perspective. *Nat Rev Drug Discov* 2004;3:763–769.

Lipner M. The polymerase chain reaction: Amplifying its role in research and beyond. *Oncol Times* 1992;14:14–16.

Lonberg N. Human antibodies from transgenic animals. *Nat Biotechnol* 2005;23:1117–1125.

Mattingly SZ. Biomarkers come of age. *Pharm Exec* 2005;25:100–111.

Melkinova I. RNA-based therapies. *Nat Rev Drug Discov* 2007;6:863–864.

Million RP. Impact of genetic diagnostics on drug development strategy. *Nat Rev Drug Discov* 2006;5:459–462.

Pack DW, Hoffman AS, Pun S, Stayton PS. Design and development of polymers for gene delivery. *Nat Rev Drug Discov* 2005;4:581.

Philippidis A. 25 up-and-coming gene therapies. *Gen Eng Biotechnol News* 2019; 39(7):16–18.

Phillips KA, Van Bebber SL. Measuring the value of pharmacogenomics. *Nat Rev Drug Discov* 2005;5:500–509.

Phillips KA, Van Bebber S, Issa AM. Diagnostics and biomarker development: Priming the pipeline. *Nat Rev Drug Discov* 2006;5:463–469.

Ratner M. FDA pharmacogenomics guidance sends a clear message to industry. *Nat Rev Drug Discov* 2005;4:359.

Roses AD. Pharmacogenetics in drug discovery and development. *Nat Rev Drug Discov* 2008;7:807–817.

Sahin U, Karikó K, Türeci Ö. mRNA-based therapeutics – Developing a new class of drugs. *Nat Rev Drug Discov* 2014;13(10):759–779.

Schilsky RL. Personalized medicine in oncology: The future is now. *Nat Rev Drug Discov* 2010;9:363–366.

Setten RL, Rossi JJ, Han S. The current state and future directions of RNAi-based therapeutics. *Nat Rev Drug Discov* 2019;18(6):421–446.

Sheridan C. Gene therapy finds its niche. *Nat Biotechnol* 2011;29:121–128.

Shih T, Vourvahis M, Singh M, Papay J. Pharmacogenetics: From bench science to the bedside. *Drug Inform J* 2008;42:503–513.

Stein CA, Castanotto D. FDA-approved oligonucleotide therapies in 2017. *Mol Ther* 2017;25(5):1069–1075.

Trusheim MR, Berndt ER, Douglas FL. Stratified medicine: Strategic and economic implications of combining drugs and clinical biomarkers. *Nat Rev Drug Discov* 2007;6:287–293.

Wang D, Tai PWL, Gao G. Adeno-associated virus vector as a platform for gene therapy delivery. *Nat Rev Drug Discov* 2019;18(5):358–378.

Chapter 7

DiMasi JA, Feldman L, Seckler A, Wilson A. Trends in risks associated with new drug development: Success rates for investigational drugs. *Clin Pharmacol Ther* 2010;87:272–277.

Evens RP. Pharma success in product development – Does biotechnology change the paradigm in product development and attrition. *AAPS J* 2016;18(1):281–285.

Evens RP, Kaitin KI. Evolution of biotechnology and its impact on healthcare. *Health Affairs* 2015;34(2):210–219.

FDA development policies and procedures. fda.gov/drugs/development-approval-process-drugs/laws-regulations-policies-and-procedures-drugs-applications.

FDA guidances. fda.gov/drugs/guidance-compliance-regulatory-information/guidance-drugs.

Lietza E, Ed. *Medical Biotechnology: Premarket and Postmarket Regulation (ebook).* ABA Book Publishing, Chicago, IL, 2015.

National Academies of Sciences. *Preparing for Future Products in Biotechnology.* National Academies Press (US), Washington, DC, 2017.

Van Norman GA. Drugs, devices and the FDA. Part 1: An overview of approval processes for drugs. *JACC* 2016;1(3):170–179.

Waller ES, Kercher NL. Laws and regulations: The discipline of regulatory affairs. In *Drug and Biological Development: From Molecule to Product and Beyond.* Evens RP, Ed., Springer Science + Business Media, New York, pp. 148–221, 2007.

Chapter 8

Anonymous. Moving up with the monoclonals. *Nat Rev Drug Discov* 2019;18(9):B5–B6.

Evens RP. *Biotechnology and Biological Products, Encyclopedia of Pharmaceutical Technology.* 2nd ed., Swarbrick J, Ed., Marcel Dekker, New York, 2001.

Evens RP. *Biotechnology and Biological Products, Encyclopedia of Pharmaceutical Technology.* 3rd ed., Swarbrick J, Ed., Marcel Dekker InForma Healthcare, New York, pp. 258–280, 2007.

Evens RP. *Biotechnology and Biological Products, Encyclopedia of Pharmaceutical Technology.* 4th ed., Swarbrick J, Ed., Taylor & Francis Group, Boca Raton, FL, vol. 1, pp. 239–266, 2013.

Evens RP. *Comprehensive Biotechnology Data Book*. Tufts University, Center for the Study of Drug Development, Boston, MA, 2020 (in press).

Evens RP, Kaitin KI. Evolution of the biotechnology and its impact on healthcare. *Health Affairs* 2015;34(2):210–219.

Lagasse HAD, Alesaki A, Simhadri VL, et al. Recent advances (therapeutic protein) drug development. *F1000Res* 2017;6(2):113. doi:10.12688/f1000research.9970.1.

Leader B, Baca QJ, Golan DE. Protein therapeutics: A summary and pharmacological classification. *Nat Rev Drug Discov* 2008;7:21–39.

Leavy O. Therapeutic antibodies: past, present and future. *Nat Rev Immunol* 2010;10:297–305.

Singh S, Kumar NK, Dwiwedi P, et al. Monoclonal antibodies review. *Curr Clin Pharmacol* 2018;13(2):85–99.

Walsh G. Biopharmaceutical benchmarks 2006. *Nat Biotechnol* 2006;24:769–736.

Walsh G. Biopharmaceutical benchmarks 2010. *Nat Biotechnol* 2010;28:917–924.

Walsh G. Biopharmaceutical benchmarks 2014. *Nat Biotechnol* 2014;32:992–1000.

Walsh G. Biopharmaceutical benchmarks 2019. *Nat Biotechnol* 2018;37:1136–1145.

Walsh G, Jeffries R. Post-translational modifications in the context of therapeutic proteins. *Nat Biotechnol* 2006;24:1241–1252.

Chapter 9

Blau HM, Daley GQ. Frontiers in medicine: Stem cells in treatment of disease. *NEJM* 2019;380:1748–1760.

Bulbake U, Doppalapudi S, Kommunineni N, Khan W. Liposomal formulations in clinical use: An updated review. *Pharmaceutics* 2017;9(12):1–33.

Evens RP. *Biotechnology and Biological Products, Encyclopedia of Pharmaceutical Technology*. 2nd ed., Swarbrick J, Ed., Marcel Deker, New York, 2001.

Evens RP. *Biotechnology and Biological Products, Encyclopedia of Pharmaceutical Technology*. 3rd ed., Swarbrick J, Ed., Marcel Dekker InForma Healthcare, New York, pp. 258–280, 2007.

Evens RP. *Biotechnology and Biological Products, Encyclopedia of Pharmaceutical Technology*. 4th ed., Swarbrick J, Ed., Taylor & Francis Group, Boca Raton, FL, vol. 1, pp. 239–266, 2013.

Evens RP. *Comprehensive Biotechnology Data Book*. Tufts University, Center for the Study of Drug Development, Boston, MA, 2020 (in press).

Evens RP, Kaitin KI. Evolution of the biotechnology and its impact on healthcare. *Health Affairs* 2015;34(2):210–219.

Feins S, Kong W, Williams EF, et al. An introduction to chimeric antigen receptor (CAR) T-cell immunotherapy for human cancer. *Am J Hematol* 2019;94(S1):S3–S9.

Fleifel D, Rahmoon MA, AlOkda A, et al. Recent advances in stem cell therapy: A focus on cancer, Parkinson's and Alzheimers. *J Gen Eng Biotechnol* 2018;16:427–432.

Fosgerau K, Hoffman T. Peptide therapeutics: Current status and future directions. *Drug Discov Today* 2015;20(1):122–128.

High KA, Roncarolo MG. Gene therapy. *NEJM* 2019;381:455–464.

June CH, Sadelain M. Chimeric antigen receptor therapy. *NEJM* 2018;379:64–73.

Keeler AM, ElMallah MK, Flotte TR. Gene therapy 2017: Progress and future directions. *Clin Transl Sci* 2017;10(4):242–248.

Lau LL, Dunn MK. Therapeutic peptides: Historical perspectives, current development trends and future directions. *Bioorg Med Chem* 2018;26:2700–2707.

Mao AS, Mooney DJ. Regenerative medicine: Current therapies and future directions. *Proc Nat Acad Sci* 2015;112(47):14452–14459.

Miliotou AN, Papadopoulou LC. CAR T-cell therapy: A new era in cancer chemotherapy. *Curr Clin Biotechnol* 2018;19(1):5–18.

Philippidis A. 25 up-and-coming gene therapies. *Gen Eng Biotechnol News* 2019;39(7):16–18.

Sahin U, Karikó K, Türeci Ö. mRNA-based therapeutics – Developing a new class of drugs. *Nat Rev Drug Discov* 2014;13(10):759–779.

Stein CA, Castanotto D. FDA-approved oligonucleotide therapies in 2017. *Mol Ther* 2017;25(5):1069–1075.

Tang J, Pearce L, O'Donnell-Tormey J, Hubbard-Lucey VM. Trends in the global immuno-oncology landscape. *Nat Rev Drug Discov* 2019;17(11):783–784.

Walsh G. Biopharmaceutical benchmarks 2006. *Nat Biotechnol* 2006;24:769–736.

Walsh G. Biopharmaceutical benchmarks 2010. *Nat Biotechnol* 2010;28:917–924.

Walsh G. Biopharmaceutical benchmarks 2014. *Nat Biotechnol* 2014;32:992–1000.

Walsh G. Biopharmaceutical benchmarks 2019. *Nat Biotechnol* 2018;37:1136–1145.

Zakrewzewski W, Dobrizynski M, Szymonowicz M, Rybak Z. Stem cells: Past, present and future. *Stem Cell Res Ther* 2019;10(68):1–22.

Chapter 10

Binder G, Bashe P. *Science Lessons: What the Business of Biotech Taught Me about Management*. Harvard Business School Press, Boston, MA, 2008.

Cohen FJ. Macro trends in pharmaceutical innovation. *Nat Rev Drug Discov* 2005;4:78–84.

Dawkes A, Papp T. Recent trends in deal-making. *Nat Rev Drug Discov* 2010;9:909.

De Rubertis F, Fleck R, Lanthaler W. Six secrets to success – How to build sustainable biotech business. *Nat Biotechnol* 2009;27:595–597.

DeFrancesco L. 2018 – Another biotech banner year. *Nat Biotechnol* 2019;37(2):116–117.

DePalma A. Twenty-five years of biotech trends. *Gen Eng News* 2005;25:1,14,16,18–20,23.

Dowden H, Munro J. Trends in clinical success rates and therapeutic focus. *Nat Rev Drug Discov* 2019;18(7):495–496.

Edwards M, Murray F, Yu R. Gold in the ivory tower: Equity rewards of outlicensing. *Nat Biotechnol* 2006;24:509–515.

Edwards MG, Murray F, Yu R. Value creation and sharing among universities, biotechnology and pharma. *Nat Biotechnol* 2003;21:618–624.

Engel S, King J. Biotech innovation. Pipeline gaps can biotech fill them? *R&D Directions* 2004;10:42–56.

Evens RP. *Comprehensive Biotechnology Data Book*. Tufts University, Center for the Study of Drug Development, Boston MA, in press, 2020.

Evens RP. Pharma success in product development – Does biotechnology change the paradigm in product development. *AAPS J* 2016;18(1):281–285.

Evens RP. The Biotechnology Industry. *J Pharm Pract* 1998;11:13–18.

Evens RP, Kaitin KI. Evolution of the biotechnology and its impact on healthcare. *Health Affairs* 2015;34(2):210–219.

Evens RP, Kaitin KI. The biotechnology innovation machine – A source of intelligent biopharmaceuticals for pharma industry – Mapping biotechnology's success. *Clin Pharmacol Therap* 2014;95(5):528–532.

Freedman Y. Locations of pharmaceutical innovation: 2000–2009. *Nat Rev Drug Discov* 2010;9:835–836.

Humphreys A, Niles S. 18th annual report. Top 100 biotechnology companies. *Med Ad News* 2009;28:6,8–12,14,16.

Micklus A, Munter S. Biopharma dealmaking in 2018. *Nat Rev Drug Discov* 2019;18(2):93–94.

Morrison C. Fresh from the biotech pipeline – 2018. *Nat Biotechnol* 2019;37(2): 118–123.

Morrison C, Lahteenmaki R. Public biotech 2018 – The numbers. *Nat Biotechnol* 2019;37(7):714–721.

Mullard A. 2018 FDA drug approvals. *Nat Rev Drug Discov* 2019;18(2):85–89.

Munos B. Lessons from 60 years of pharmaceutical innovation. *Nat Rev Drug Discov* 2009;8(12):959–968.

Niles S. 19th annual report. Top 100 biotechnology companies. A mixed bag for biotech. *Med Ad News* 2010;29:1,6,8,10,12,14,16,18–19.

Paul SM, Mytelka DS, Dunwiddie CT, et al. How to improve R&D productivity: The pharmaceutical industry's grand challenge. *Nat Rev Drug Discov* 2010;9(3):203–214.

Pavlov AK, Belksey MJ. BioPharma licensing and M&A trends. *Nat Rev Drug Discov* 2005;4:273–274.

Report Medicines in Development. Biotechnology. Pharmaceutical Research and Manufacturers Association, Washington, DC, 2008.

Scannell J, Blanckley A, Boldon H, Warrington B. Diagnosing the decline in pharmaceutical R&D efficiency. *Nat Rev Drug Discov* 2012;11(3):191–200.

Index

Note: Page numbers in italic and bold refer to figures and tables, respectively.